An Architecture of Parts

This book is unique in describing the history of post-war reconstruction in Britain from an entirely new perspective by focusing on the changing relationship between architects and building workers. It considers individual, as well as collective, interactions with technical change and in doing so brings together, for the first time, an extraordinary range of sources including technical archives, oral history and visual material to describe the construction process both during and in the decades after the war. It focuses on the social aspects of production and the changes in working life for architects and building workers with increasing industrialisation, in particular, analysing the effect on the building process of introducing dimensionally co-ordinated components.

Both architects and building workers have been accused of creating a built environment now popularly discredited: architects responsible for poor design and building workers for poor workmanship. However, many of the structures and ideas underpinning this period of rapid change were revolutionary in their commitment to a complete transformation of the building process. *An Architecture of Parts* adds to the growing literature on changes in the building world during and immediately after the Second World War. It is significant, both empirically and historically, in its examination of the ideas, technology and relationships that fired industrialisation of the building process in mid-century Britain.

Christine Wall is a Senior Research Fellow at the Centre for the Study of the Production of the Built Environment (ProBE), School of Architecture and the Built Environment, University of Westminster.

Routledge Research in Architecture

The *Routledge Research in Architecture* series provides the reader with the latest scholarship in the field of architecture. The series publishes research from across the globe and covers areas as diverse as architectural history and theory, technology, digital architecture, structures, materials, details, design, monographs of architects, interior design and much more. By making these studies available to the worldwide academic community, the series aims to promote quality architectural research.

An Architecture of Parts: Architects, Building Workers
and Industrialisation in Britain 1940–1970
Christine Wall

An Architecture of Parts

Architects, building workers
and industrialisation in Britain
1940–1970

Christine Wall

Routledge
Taylor & Francis Group

LONDON AND NEW YORK

First published 2013
by Routledge
2 Park Square, Milton Park, Abingdon, Oxfordshire OX14 4RN

Simultaneously published in the USA and Canada
by Routledge
711 Third Avenue, New York, NY 10017

First issued in paperback 2016

Routledge is an imprint of the Taylor & Francis Group, an informa business

© 2013, Christine Wall

The right of Christine Wall to be identified as author of this work has been asserted by her in accordance with sections 77 and 78 of the Copyright, Designs and Patents Act 1988.

All rights reserved. No part of this book may be reprinted or reproduced or utilised in any form or by any electronic, mechanical, or other means, now known or hereafter invented, including photocopying and recording, or in any information storage or retrieval system, without permission in writing from the publishers.

Trademark notice: Product or corporate names may be trademarks or registered trademarks, and are used only for identification and explanation without intent to infringe.

British Library Cataloguing in Publication Data
A catalogue record for this book is available from the British Library

Library of Congress Cataloging-in-Publication Data
Wall, Christine.
An architecture of parts : architects, building workers and industrialization in Britain 1940-1970 / Christine Wall.
pages cm – (Routledge research in architecture)
Includes bibliographical references and index.
1. Architects and builders–Great Britain–History–20th century.
2. Architecture and society–Great Britain–History–20th century.
3. Construction industry–Social aspects–Great Britain. 4. Industrial sociology–Great Britain. 5. Reconstruction (1939–1951)–Great Britain.
I. Title.

NA2543.B84W34 2013
720.941'09045–dc23
2013003440

ISBN 13: 978-1-138-22935-8 (pbk)
ISBN 13: 978-0-415-63794-7 (hbk)

Typeset in Sabon
by Keystroke, Station Road, Codsall, Wolverhampton

Dedicated to Douglas S. Wall (1924–2012)

Contents

List of figures	ix
List of tables	xi
Acknowledgements	xiii
List of abbreviations	xv

Introduction	1

PART I
Industrialisation and the British building industry — 15

1 The industrialisation of building	17
2 The building industry during war and reconstruction	32
3 Education and training	51
4 Post-war change: management and organisation	61

PART II
Architectural abstraction: the role of the Modular Society in promoting industrialised methods — 77

5 The Modular Society	79
6 'Additive architecture': the early years of modular co-ordination	95
7 The BRS and the mathematisation of architectural modularity	116

PART III
'Never argue with the architect': architects and building workers 1940–70 — 133

8 'Put nobody between the architect and the men': the role of architects on site — 135

9 The nature of work in the construction industry — 154

10 Elusive connections: architects and building workers in mid-century Britain — 171

Appendices — 184
Glossary of modular terms — 188
Notes — 189
Bibliography — 209
Illustration credits — 225
Index — 227

Figures

1.1 Cover of *New Ways of Building* (1948)	18
1.2 Trade advertisement for system building, early 1960s	24
2.1 Building workers, c.1943, listening to a Colonel returned from the Malta campaign	38
2.2 Building workers, c. 1943, in conversation with a Colonel returned from the Malta campaign	39
2.3 Woman bricklayer in the Second World War	39
4.1 Modern plant for brickwork	68
5.1 Cover of The Modular Society flyer, c. mid-1960s	80
5.2 Lord Bossom examining a Modular Society questionnaire while on a site visit	85
6.1 Cover of *The Modular Quarterly* showing the first Modular Assembly	104
6.2 Presentation drawings for the first Modular Assembly	105
6.3 Mark Hartland Thomas on the steps of the first Modular Assembly	106
6.4 The five essentials of modular co-ordination, a summary of five years' work	110
6.5 The second Modular Assembly at IBSAC 1964	112
6.6 The second Modular Assembly exhibition of modular components	113
6.7 The third Modular Assembly at IBSAC 1966	114
6.8 Modular Society members at the Society's conference, 18 November 1963. From left to right, Alan Diprose (Modular Primer), P.H. Dunstone (Combinations), Mark Hartland Thomas (Secretary), Peter Trench (Chair), Bruce Martin (International Work)	115
7.1 Three-dimensional model of Ehrenkrantz's number pattern found inside the back cover of *The Modular Number Pattern* (1956)	120
7.2 Photo of British Standards Institute test buildings, EPA Project, designed by Bruce Martin	126

x *List of figures*

7.3 Scale drawing of British Standards Institute test buildings, EPA Project, designed by Bruce Martin	127
7.4 Building Research Station test buildings, EPA Project. Rationalised traditional terraced housing in brick and block	128
7.5 Plans of BRS terraced housing in brick and block	129
8.1 Little Aden Cantonment, Farmer and Dark, 1961. Photographs of site model	149
8.2 Little Aden Cantonment, plans and elevations of officers' housing	151
8.3 Little Aden Cantonment, plans and elevations of sergeants' mess	152
10.1 Ratings of 'contribution to the building process' and 'social status' by occupation	180
10.2 A75 Metric building under construction	183

Tables

1.1	Industrialised dwellings: local authorities and new towns in England and Wales	25
2.1	Structure of the construction industry by size of firm, 1935–53	34
4.1	Proportion of unskilled in the construction workforce, 1946–54 (private sector firms only)	72
4.2	Proportion of unskilled in the construction workforce, 1958–70 (private sector firms only)	72
9.1	Women architects newly admitted to ARCUK register, 1960–69	162
9.2	Comparative professional earnings in 1955 (from the Pilkington Inquiry)	164
A.1	Visits by Modular Society members to modular buildings, 1953–56	184

Acknowledgements

Two seminal texts of twentieth-century architectural and construction history, Linda Clarke's *Building Capitalism* (1992) and Andrew Saint's *Towards a Social Architecture* (1987) provided the inspiration for this book. Andrew Saint, my PhD supervisor at the Department of Architecture, University of Cambridge, introduced me to many of the people I interviewed: his support of the original research was invaluable. This book could not have been written, however, without the hospitality and generosity of the many architects who spent considerable time talking to me and explaining their positions on industrialisation, modularity, and their role in the construction process. I would like to thank in particular Alan Crocker, Sir Andrew Derbyshire, Alan Meikle, Steve Mullin, Nick Woodhouse and the late Sir Roger Walters. I am especially indebted to Bruce Martin who gave me unlimited access to his papers on the Modular Society and EPA Project 174, and invited me to work in his studio.

I would also like to thank former building workers Don Baldrey and Peter Barnes, who gave graphic accounts of building with CLASP and working life on site in the 1950s and 1960s, and Mike Hatchett for sending a detailed critique of the impact of industrialisation on site work.

I am very grateful to Linda Clarke for providing support over many years, introducing me to Donald Bishop, and also for reading and commenting on the Introduction. I would also like to thank Nick Bullock, Peter Carolin, and Richard Hill for their positive comments and encouragement of the original thesis. More recently, Philip Steadman and Andrew Rabenack made helpful comments on the Modular Society chapters.

The original research was supported by an RIBA LKE Ozolins studentship, and the School of Architecture and the Built Environment, University of Westminster, under Jeremy Till, provided a grant to cover the costs of illustrations. I would also like to thank Fran Ford and the team at Routledge for bringing this book into production.

Publication of this book has been supported by the Centre for the Study of the Production of the Built Environment (ProBE), the University of Westminster.

I would like to thank the publishers for permission to reproduce parts of Chapter 2 which previously appeared, in a slightly different form, in 'The development of building labour in Britain', in A. Dainty (ed.) *HRM in Construction: Critical Perspectives*, London: Taylor & Francis (with L. Clarke and C. McGuire).

Permission to reproduce the illustrations in the text was kindly given by the following, Architectural Press Figure 1.1; Stramit Technology Holdings Limited, Figure 1.2; Imperial War Museum Photograph Archive Figures 2.1, 2.2, 2.3; RIBA Photograph Collection, Figure 6.3; the OECD Figures 7.2, 7.3, 7.4, 7.5. All other illustrative material is reproduced from the *Modular Quarterly*, the journal of a private society now disbanded.

Abbreviations

AA	Architectural Association
AACP	Anglo-American Council on Productivity
AASTA	Association of Architects, Surveyors and Technical Assistants
ABT	Association of Building Technicians
ARCUK	Architect's Registration Council UK
ASB	Architectural Science Board
ASW	Amalgamated Society of Woodworkers
AUBTW	Amalgamated Union of Building Trade Workers
BATC	Building Apprenticeship and Training Council
BICRP	Building Industry Communications Research Project
BINC	Building Industry National Council
BRS	Building Research Station
BS	British Standards
BSI	British Standards Institute
CEU	Constructional Engineering Union
CITB	Construction Industry Training Board
CLASP	Consortium of Local Authorities Special Programme
COID	Council of Industrial Design
dc	dimensional co-ordination
DIN	Deutsches Institut für Normung (German Institute for Standardisation)
DLO	Direct Labour Organisation
ECA	European Co-operation Administration
EPA	European Productivity Agency
EWO	Essential Work Order
GTC	Government Training Centre
IB	industrialised building
ICE	Institute of Civil Engineers
ISE	Institute of Structural Engineers
ITB	Industrial Training Board
LCC	London County Council
MHLG	Ministry of Housing and Local Government
MoW	Ministry of Works

NFBTE	National Federation of Building Trade Employers
NFBTO	National Federation of Building Trade Operatives
NJCBI	National Joint Council for the Building Industry
NPOT	New Pattern of Operative Training
OECD	Organisation for Economic Co-operation and Development
OEEC	Organisation for European Economic Co-operation (superseded by OECD in 1961)
PBR	payment by results
RIBA	Royal Institute of British Architects
RICS	Royal Institute of Chartered Surveyors
RSM	Research into Site Management
SMM	Standard Method of Measurement
VET	vocational education and training

Archives

MRC	Modern Records Centre
PRO	Public Records Office
RIBA	Royal Institute of British Architects
TNA	The National Archives

Journals

AD	*Architectural Design*
AJ	*Architects' Journal*
IBSAC	*Industrialised Building Systems and Components*
MQ	*Modular Quarterly*
NBL	*New Builder's Leader*
OAP	*Official Architecture and Planning*
RIBAJ	*Journal of the Royal Institute of British Architects*

Introduction

The reconstruction of Britain at the end of the Second World War demanded a rapid increase in output from a construction industry suffering severe shortages of labour and materials. The solution proposed was the industrialisation of production, predicated on the rationalisation of design and construction processes, and the use of non-traditional materials and methods, which had been thoroughly researched during the 1930s. These, together with dimensional co-ordination of building components and modular design, were among the innovations promoted throughout the 1950s and 1960s as key factors in increasing the efficiency of construction. They also produced an identifiable, modern aesthetic – although it can be argued that this, in many cases, was an unintended consequence of the techniques used.

This book describes the key transformations in the mid-twentieth-century building world as industrialisation took hold and 'process' rather than 'product' fired the imagination and dominated debates on construction, for both architects and building workers. This period saw transformations in the structure of the building industry in terms of size of firms and forms of employment; in education and training as barriers between vocational and architectural training increased; in management and organisation as industrialised building became prevalent; in the skills and occupational structure of the workforce, and in the increasing invisibility of the manual construction workforce in discourses on the production of the built environment.

It focuses on the social context for production at a time of rapid technical change: the social structure of the construction industry, its gendered composition, its social stratification and the different routes for acquiring specific education, training and skills are examined in depth. This reveals the relationships, interactions and increasing distance between architects and building workers over three decades. Industrialisation threatened the status of both architects and building workers and both groups experienced far-reaching changes in their working lives and education and training. This book provides an account of the introduction of a new system of production, one that transformed the traditional role of the skilled worker and also that of the architect.

2 Introduction

The book is structured into three parts: Part I introduces the ideas underpinning industrialisation and the social context for their implementation in the British building industry. Part II presents an in-depth analysis of the work of the Modular Society from 1953 onwards, and its campaign for a universal system of modular co-ordination. This was considered in many quarters to be fundamental for the successful industrialisation of the construction industry. The concept of modular component building was essentially abstract, originating from an engineering and mathematical base. It assumed the imposition of order and sequence on the building process and though this implies a change to the social process of building to accompany the precision of a minutely planned exercise, site processes were often ignored. Part III describes the nature of work for architects and building workers in relation to industrialised methods of building and contextualises these in contemporary and historic accounts of the period. Finally, Chapter 10 draws together these three distinct strands in a new appraisal of the post-war period of reconstruction.

Technical change in the construction industry

Innovation, defined as the application of invention in order to increase efficiency and ultimately profit, is an essential factor in the process of industrialisation. There are remarkably few texts that focus on innovation in the construction industry; in fact, it is notable by its absence from economic histories. This makes the achievement of Marian Bowley, in her comprehensive account of the British building industry published in 1966, especially noteworthy. In *The British Building Industry: Four Studies in Response and Resistance to Change* (1966), Bowley describes the difficulties in introducing innovations in both structures and methods of building and design. She notes that by the end of the Second World War, technical knowledge in construction had outstripped architectural and building practice. Despite a large 'stock' of technical innovations in both materials and components for building acquired from both government-funded and private business research during the inter-war years, there was very slow uptake in the building process.

She then proceeds to analyse 'the system' by which buildings are produced and finds that the separation of design from production is a serious defect in organisation. She traces the historical origins of this separation culminating in the 1931 Architects Registration Act, which prevented members of the profession from holding a profit-making position in the building industry. The consequence of this was that builders' sons were trained as engineers in order that they could join the family firm, as, unlike architects, professional engineers had no such exclusionary clause in their constitution. Architects became more closely aligned with the client and, therefore, distant from the construction industry. But contract norms still gave architects the powerful position of supervising the whole building process after creating designs that were first of all handed to engineers for structural detailing and then to surveyors for costing. Bowley, throughout her account, accuses the architectural profession of being the weak link in the chain of production and innovation due to architects' lack of technical knowledge and ignorance of costing, resulting in designs of no 'economic rationality'. She concludes:

> [T]he traditional organisation of the profession is not appropriate for providing the design services required today ... It might be desirable to integrate into one profession all the disciplines which are required for the design of buildings and their production, establishing a common basic education and varying degrees of specialisation.[1]

Bowley does not consider the training or skills of the workforce to be a factor in the efficiency of building. Here, labour is regarded as a commodity and is discussed only in relation to scarcity or surplus. The consequences of full employment in the 1950s are described in terms of effective site organisation, so that labour is utilised continuously in order to save costs. To maximise efficiency, she suggests the introduction of full preparation of design drawings before work starts, components that do not have to be adjusted on site, more pre-assembly of components and dimensional co-ordination. Bowley's history describes an industry held tightly together by a rigid and institutionalised establishment that is incapable of change. She compares it to a complicated mosaic of ill-fitting relationships between designers and builders and considers improvement impossible until the pattern of relationships becomes more suitable for efficient building.

David Gann (2000) has defended the construction industry as 'innately innovative' as opposed to Marian Bowley's description of the industry as a passive recipient of innovations from other sectors. However, he still depicts it as an industry suffering from serious structural deficiencies described as 'ghosts' haunting the industry, acting as barriers to innovation. He argues that when innovation does take place, it occurs within a closed loop inside the framework of an individual project so that only marginal benefits diffuse out to the wider industry.[2] Gann includes only a limited analysis of innovation in relation to social processes and structures, but both he and Bowley present an industry fractured by the often opposing interests of users, designers and builders and constrained by the poor communication between these actors.

The slowness of the construction industry to apply innovations can also be interpreted as part of a wider problem inherent in the education and training of engineers and architects in Britain. This argument is put forward by those typified by David Edgerton,[3] as 'declinist' historians of technology, for example, Barnett (1986) and Wiener (1981), who blame Britain's perceived poor economic performance since the war on low levels of specialist technical education, especially in comparison with Germany.[4] This theme was reiterated by Landes (1998) who describes the British provision of technical education as lacking in clear policy and having been allowed to 'grow like a weed and . . . once grown [treated], like a poor relation of "proper" schools and universities'.[5] The declinist view also includes the poor funding of research and development in British industry, by both government and within firms, as a factor that resulted in lack of innovation (Landes, 1969; Mowery, 1986). This declinist view is challenged by David Edgerton, who argues in *Warfare State: Britain 1920–1970* (2006), that Britain supported a growing and dynamic armaments industry as well as developing a welfare state. This is exemplified by Donald Gibson's appointment as Chief Architect to the War Office in 1957, where he directed a research department, which developed the NENK method, an industrialised system for military dwellings.

Michael Ball (1988), however, considered construction as having characteristics that made it unique in comparison with other industries, whereby the division between design and production is an outcome of the social divisions of labour imposed by the contracting system.[6] Innovation in technique is limited under this system by a number of factors. First, there is the fragmentation of the building process between a number of agents: designers, surveyors, contractors, plant-hire firms and materials suppliers. Second, there is the nature of the competition created by the competitive tendering procedure won on the basis of lowest cost. Third, Ball cites the exploitation of labour where discontinuity arises from laying off workers in time of slack, disbanding of a team at the end of a project and extensive subcontracting and self-employment. All these factors lead to poor transmission of knowledge within the industry.

Ball also describes the role the state takes in legalising the framework for the contracting system and as a major client in the post-war years of the Welfare State and notes the strong alliance between contractors, the state, and the professions against the demands of the workforce. Building workers were seen, even under Harold Wilson's Labour governments as 'a problem of the building industry'.[7] Ball demonstrates that, in the post-war years, construction cycles were no different from other manufacturing industries; in fact, there was comparatively less variation in output. By contrast, construction showed very much higher variation in employment than other sectors of the economy so that 'the success of capital was won at considerable expense to the workforce'.[8]

Economic histories of the construction industry in the twentieth century, for example, Hillebrandt (1984), and Powell (1980), take a largely neo-classical view so that the developments and dynamics of training, education and social relationships are not analysed. Marxist perspectives, explaining change in terms of crisis, revolution and discontinuity, provide richer sources for analysis through a theoretical breadth that includes political events, and the interdependence of economic and social development.[9] An extensive collection of writings exemplifying this approach can be found in the collection of the Bartlett International Summer School (BISS) Proceedings.[10]

Introduction 5

This has recently been reviewed (Clarke and Janssen, 2008) and assessed as arising in response to particular political developments of the time, but many of the debates and assumptions, in particular, the centrality of the labour process to production, are still important as a way of understanding built form and how it expresses the condition under which it was constructed.

The post-war era of industrialised production of housing has been documented from the perspective of housing policy (Donnison, 1960; Cullingworth, 1966), political economy (Merrett, 1979), social history (Burnett, 1986), and technological change (Russell, 1981; Sebestyén, 1998). Of those accounts that take a technological standpoint, one in particular stands out by including elements of both the political and social context for innovation: R.B. White's *Prefabrication: A History of its Development in Great Britain* (Building Research Station, 1965). Here, the prefabrication movement of the 1920s and 1930s is regarded as a precursor to industrialisation and the inter-war period as a time of innovation in both materials and techniques. White agrees with Bowley in noting the extensive research into non-traditional methods of building existing in Britain and the bulk of the book is taken up with meticulously detailed and illustrated accounts of these experiments and summaries of the government reports that affected construction. He acknowledges that, by the 1950s, other Western European countries were leading in their use of tower cranes and industrialised systems for multi-storey building but, unlike Bowley, does not blame this late uptake of mechanisation in Britain on the structure of the construction industry but on government policy. White argues that post-war policy was not wholly committed to industrialisation, being based instead on 'expediency and emergency measures' without any long-term planning for industrialisation. This lack of commitment was based on the assumption that traditional methods would be returned to when the numbers in the construction workforce had been increased. This dual approach resulted in:

> a dispersal of effort over the whole field of research into alternative methods ... The reversion to traditional or 'improved traditional' methods in house building ... was nevertheless, rapid enough to inhibit the further development of those non-traditional systems that had attained some temporary measure of success.[11]

While agreeing that prefabrication is commonly introduced to combat skill shortages by shifting more of the labour input into the factory rather than on the site, White does not subscribe to the view that site assembly only requires semi-skilled labour, 'in non-traditional building the new skills are often said to be generally better understood and performed by those who have had a basic training in some traditional craft or process'.[12] The school building programme is taken as an exemplar and White concludes optimistically that prefabrication and industrialisation are inevitable but that there is no reason for this to be incompatible with 'a better visual background'.

The first major text to undertake an examination of the political and organisational forces behind the production of industrialised dwellings was that of Dunleavy in 1981. In *The Politics of Mass Housing in Britain 1945–75*, he argues that high-rise flats were built mainly because of construction industry pressure from large contracting firms. He cites contractor-funded feasibility studies, industrial advertising (using examples from the *Municipal Journal* aimed at local government personnel) and a high-profile industrialised building campaign based on the success of the school building programme, together with the availability of heavy prefabricated systems from France and Scandinavia.[13] Other factors were the link between high-rise and the architectural ideology of the Modern Movement, the popular association of high-rise with technical progress and erroneous planning methodology that argued in favour of high-density urban high-rise in order to preserve scarce and expensive land.[14] In this account, construction labour does not appear to exert any influence on either the introduction or the demise of high-rise buildings and industrialised processes.

Brian Finnimore (1989), however, situates his account of factory-built, post-war housing firmly within the economic framework of the Welfare State and considers the roles of both architects and operatives. He traces the initial resistance of building craft unions, represented by the National Federation of Building Trades Operatives (NFBTO), to industrialisation in the 1950s, giving way to acceptance by the mid-1960s. The trade union movement is presented as being in an ambiguous position. Strong propaganda from the government on the drive for public-housing renewal made union leaders sensitive to public opinion and they did not want to betray class loyalties by being seen to impede the slum clearance programme. At the same time the membership was wary of the introduction of industrial methods, as they were feared by many to result in a loss of craft skills.[15]

Although Finnimore notes the increase in semi-skilled workers during this period of industrialisation, particularly the period of heavy pre-cast concrete panel systems, he does not relate this to the question of appropriate training and skills for concrete work. He does, however, state the neo-classical economic argument on productivity: that the cheapness of British building labour inhibited attempts to fully industrialise house production and that economies would only be realised, in comparison to traditional house building, if wages were higher.

The publication of *Tower Block* in 1994 by Miles Glendinning and Stephan Muthesius gave a new perspective on the same phenomenon.[16] Whereas the two previous authors are united in their interpretation of industrialised high-rise housing as a failure of national housing policy and positively damaging to the working-class tenants forced to live in the flats, Glendinning and Muthesius set out to convince that high-rise housing was not entirely an unqualified disaster. They focus on the local framework that enabled high building to progress with such rapidity in the short years in which it proliferated and consider the main agents to have been the designers and 'producers'. Here 'producers' are defined as the councillors and supporting officials who provided the political and organisational impetus behind large-scale production.

In a complete reversal of Dunleavy's findings, they assert that local government officials, both architects and engineers, acted to assist councillors and that it was 'frequently their duty to cajole the contractors into offering large package deal contracts for high blocks'.[17] Although a large part of the book concerns production, the definition of 'producers' used precludes any examination of the role of labour except for that of the local authority Direct Labour Organisations (DLOs). These are mentioned intermittently, and usually unfavourably, as the only form of organised building labour to have any direct effect on what was built.[18] Despite this, the authors are aware that the 'political and architectural rhetoric of mass-production and system-building was curiously detached from the day-to-day organisation of building.'[19] They do not consider that industrialisation led to de-skilling of the labour force and indeed comment on its beneficial effects on site organisation. An engineer who worked on one of the first tower crane sites, a London County Council development in 1953, is quoted as describing the site as having 'complete order and an absence of workmen in contrast to the traditional over-populated, semi-organised rubbish dump'.[20]

The most comprehensive architectural account of the era of industrialised building is Russell's 1981 survey of systems building.[21] This starts with the traditional Japanese house and takes in the late nineteenth and late twentieth century and, while centrally preoccupied with viewing industrialisation through an architectural lens, includes at every stage of development an analysis of the role of labour. This is refreshing and unusual in an architectural history and provides interesting material on the types of skill required for industrialised building. Russell claims that the building industry in the 1980s was still craft-based and that arguments for industrialisation, made by contractors and architects and based on skill shortages, were selective. For this he cites the CLASP school building programme, where skill shortages only arose sporadically and were not constant:

> Many of the schools consortia found, not that there were shortages of craftsmen, but that often there were complaints that contractors could not use the craft skills which were available but had to learn new ones to go with the nut and bolt technology.
> (Russell, 1981, p. 318)

The concept of skill in the construction industry

It is impossible to consider the issue of skill without referring to the labour process, a theoretical concept originating from Marx's theory of surplus value, summarised by Linda Clarke as 'the process whereby labour is appropriated, subsumed and consumed within the production process'.[22] The only significant text to apply this to the construction industry is Linda Clarke's (1992) *Building Capitalism*, which focuses on the building labour process in the development of Brill Farm, later to become Somers Town, in eighteenth-century London. Here, it is argued that the particular form of urbanisation that arose in this district was dependent on the changing conditions and relations of production between building labour, contractors and land-owners. In effect, Clarke argues that the built environment was transformed in time, in a manner specific to a particular place, by the production process.

For building labour, the definition and recognition of skill are of paramount importance in negotiating wages, the content, form and control of training and in forming individual and collective identity. The usual British notion of skill is conceived as the ability, usually physical, to fulfil a particular task or activity in the workplace in order to produce a given output. It can be traced from the eighteenth century, from Adam Smith through Ruskin to Braverman and is a peculiarly Anglo-Saxon notion, used in a similar way to 'know-how' and 'technique', the worker with 'skill' being understood to possess know-how appropriate to the task in hand (Clarke and Winch, 2006).[23] The earliest formulation of Ruskin's interest in the process of building is found in *The Seven Lamps of Architecture* published in 1849: 'I believe the right question to ask, respecting all ornament, is simply this: was it done with enjoyment – was the carver happy while he was about it?'[24] This is probably the most often quoted and frequently misquoted, reference by Ruskin to the emotional state of the craftsman while at work. It was written at a time when the Gothic revival was in full swing and partly in response to Pugin's books, *Contrasts* (1836) and *The True Principles of Pointed or Christian Architecture*, published in 1841, in which he argued that the Catholic architecture of the late thirteenth century should be taken as exemplary. The most familiar example of this architecture is Barry and Pugin's Houses of Parliament, started in 1837 and finished in 1860. Ruskin answered Pugin from a different religious standpoint and through a meticulous study of the early Gothic (1200) cathedrals of Salisbury and Normandy, which he illustrated with careful measured drawings, published as engravings in *Seven Lamps*.

This interest in the well-being of the craft worker, and the relations between architects, builders and workers was a theme that recurred throughout Ruskin's life. Years later, in his *Fors Clavigera* letters to the "working men and labourers" of England, Ruskin dispensed with his usual publisher in an attempt to communicate directly with his readership through a series of monthly pamphlets setting out the means to put his ideals into practice.[25]

This emphasis on happiness was more than a straightforward plea for ethical and just conditions of employment. Ruskin conceived buildings as a testament to the quality of human labour expended on them and thus as an expression of the human soul in material form. In other words, he believed that the quality of architecture was directly affected by the state of the worker who produced it – but as Ruskin's definition of what constituted architecture was limited to buildings with decoration/ornamentation, he was referring initially not to all building workers but in particular to ornamental masons. He extended his ideas on this subject in the sixth chapter of the second volume of the *Stones of Venice* (1853) entitled 'On the Nature of Gothic Architecture' which became an influential tract. Here, Ruskin proposes that true Gothic architecture is characterised by the combination of its elements; which he explains as the mental tendencies of the builders and its forms – the pointed arches, vaulted roofs and other architectural characteristics. Ruskin elevates work produced by free Christian men over the classical architecture of Greece, which was the work of slaves. Even if freedom results in imperfection and mistakes, at least it is the outcome of a man at work not a machine or man toiling as an 'animated tool'. He argues that, 'Men were not intended to work with the accuracy of tools, to be precise and perfect in their actions',[26] and equates imperfection with good architecture, asserting that precision and accuracy of workmanship are more often a sign of poor architecture and rarely a sign of the best.

Ruskin explains the revolutionary, political uprisings of the working classes in the 1840s as due not to low wages, starvation and famine but because of 'the degradation of the operative to a machine'. This results in a 'lack of pleasure' in work and the loss of the working man's soul with the division of labour into repetitive tasks of factory work. He encourages the middle classes not to buy mass-produced goods because of the poor working conditions implicit in their manufacture: an early plea for ethical consumption. He also argues against the separation of intellectual and manual work saying; 'it would be well if all of us were good handicraftsman in some kind, and the dishonour of manual labour done away with altogether.' Following on with the famous phrase that inspired so many Arts and Crafts architects; 'The painter should grind his own colours; the architect work in the mason's yard with his men.'[27]

Ruskin has been criticised for being historically inaccurate, especially in his descriptions of medieval working practices, but there were contemporary reasons for his romanticisation of medieval working life. Ruskin's elevation of it can be seen as a response to the form of employment commonplace in the manufacturing industries of England by the mid-nineteenth century, in particular, the division of labour and the miserable conditions found in many small sweatshops.

Brian Hanson (2003) has suggested that Ruskin's ideas were reinforced through contact with George Gilbert Scott who recounted his visits to Germany in the 1840s. German building processes constituted a social microcosm highly compatible with Ruskin's amalgam of the artistic and the political. A good example of this was the restoration of Cologne Cathedral, started in 1820s, and called for by Schinkel as necessary as:

> a monument whose method of construction would implant in the nation something that would bear fruit . . . recalling the old craftsmanship ethic

of our forefathers, in which love, humility and a just sense of pride combined . . . [to produce] a living and regenerative monument.[28]

Adjacent to the cathedral throughout its restoration were the *Bauhütte*, the workshops and school for apprentice craft workers, where Scott had observed a much less hierarchical relationship between the intellectual and operative parts of the enterprise than he expected.

So Ruskin's view of the relationship between architect and building worker – designing mind and executive hand – was not entirely sentimental talk of happy carvers. He argued for a close relationship between the architect and the craftsmen, preferably with the architect gaining some hands-on experience of the practicalities of building. But the opportunities for this type of collaboration were fast disappearing in the urban expansion of the late nineteenth century, alongside changes in building contracting and the professionalisation of the architect's role.

The conception of the skilled worker enriched through Ruskin's depiction of the skilled, masculine worker, embodying independence, creativity and autonomy, rapidly became part of the discourse on work for both employers and trade unions.[29] His rhetorical descriptions of physical labour used as part of his argument against the separation of intellectual and manual work came into mass circulation later when reprinted in pamphlet form by William Morris.[30] *On the Nature of Gothic* became a manifesto for the Arts and Crafts Movement and a founding text for F. D. Maurice's, Working Men's College. Morris re-interpreted Ruskin in a form far less damning of technology and the modern world than those of his mentor and evident in his popular lecture *Useful Work Versus Useless Toil*, published as a Socialist League pamphlet.[31] The nineteenth-century craft trade unions wholeheartedly assumed Morris' and Ruskin's portrayals of the noble, male, skilled worker who was made graphically explicit in many of the union banners. For example, the emblem of the Amalgamated Society of Carpenters and Joiners in 1866 depicts the carpenter Joseph of Nazareth, together with white-aproned joiners, surrounded by the draped allegorical female figures of Industry and Art, Justice and Truth. The figure of the 'free', 'happy' and autonomous craftsman and Ruskin's idealisation of manual work have had a lasting, if somewhat ambiguous, legacy well into the twentieth century, particularly so in the construction industry, and have contributed to a specific understanding of skill as an individual attribute measured by individual output in the product.

Braverman's influential, but now superseded, Marxist analysis of the labour process in relation to industrialisation was founded on this interpretation. He argued that increased mechanisation would lead to a loss of skill in the workforce due to 'de-skilling' through the fragmentation of tasks, and decreasing knowledge required for their implementation, associated with assembly line production.[32] This again relied on narrow definitions of skill and craft as attributes of individual workers so that skill was regarded as physically rather than socially determined. Again, this definition coincides with the attributes of the traditional apprentice: learning largely on the job, where possible with a journeyman, with little theoretical underpinning and often fairly unsystematically.[33] The same notion has recently been resurrected by Richard Sennett in his aptly titled book *The Craftsman*.[34]

There are difficulties in this conception of skill; a key problem is that skill is defined in the tasks or outputs of the individual worker in the workplace rather than in the division of labour at a collective and social level, between, for example, carpenters and bricklayers (Clarke and Janssen, 2008: 3). It is therefore equated with a particular conception of labour as a commodity, with what Richard Biernacki terms 'embodied labour' as distinct from the abstract power of labour which he terms 'labour power'.[35] The reference point for this Anglo-Saxon concept is not the realisation of labour potential but the *object* of labour, bound to traditional tasks and activities, functional in nature and defined by an individual performance. As will be shown later, the importance of this concept is that it is separate from the process of vocational education and training (VET), which has the potential to integrate knowledge with personal, social and technical competences and adapt to new technologies and the changing organisation of work (Brockmann et al., 2008).[36] However, it underpinned much of the research on productivity and new methods of building undertaken at the Building Research Station (BRS), the many government reports on the industry, and contributed to the archaic management practices prevalent throughout the construction industry.

In *The Fabrication of Labor*, Richard Biernacki, using Marx's concept of surplus value as a basis for analysis, compares the development of the wool textile industries in Britain and Germany between 1640 and 1914, through an examination of the micro-relations of production on the factory floor. These are considered the key factors in the development of the respective industries, and contribute to different, nationally specific understandings of the nature of labour as a commodity, which underpinned the divergent paths of their respective development in Britain and Germany.

Although this account is centuries away from the period to be studied here, concerns different industrial sectors and describes the working of liberal capitalist economies before the intervention of state regulation, it has potential as a theoretical tool for explaining the nature of productive relations, on sites, in the British construction industry.

The primary justification for using this framework must be that the British construction industry was remarkably free of state intervention in the 1950s and 1960s. As will be shown, unlike other industries of similar capacity and organisation, it was not nationalised by the post-war Labour government. As such, the buying and selling of labour power became an important node around which productive relations, and from there the nature of the industry, developed. To explore the relevance of this framework further, it is worth elaborating the core findings in Biernacki's work. Through a meticulous, comparative analysis of systems for measuring cloth produced, and hence the wage paid, he reveals the exact nature of the transaction between employer and worker. In Britain, labour power was appropriated through the product. Hence the method of wage payment, piece-rates, was calculated on the basis of 'hours per cloth' and workers were paid a sum per length and density of cloth. Workers, however, retained ownership of their labour power and this was reflected in the culture of the industry; weavers looking for work asked an employer if he had any 'looms to let' and workers did not change out of their own clothes into factory overalls on starting work.[37]

Biernacki argues that the planning and architecture of British mills arose from this conception of labour. Crossing the threshold from the outside to the inside became important as the point of subordination of the worker and thus the significance of the main entrance was emphasised by flanking turrets and ornamentation. Factory layout in nineteenth-century Britain was that of a fortress-like enclosure around a mill yard with a single main entrance gate. The managers exerted rigid controls on the entry and exit of workers through the gate at times gauged to the second by the mill clock, and latecomers were invariably locked out.

By contrast, the architecture of German mills did not emphasise the threshold but instead, the movement of workers was constricted once they were inside the mill. There was no central yard but a series of interconnected rooms for different functions, each with an elaborate system of lobbies that prevented free movement of workers within the factory. The emphasis was on the worker remaining with his or her machine under the gaze of the 'overlooker'. Biernacki argues that this spatial arrangement arose from a conception of the appropriation of labour as 'the continuous transformation in time of labour power into a product'.[38] Hence the piece-rates were calculated as 'cloth per hour' and workers were paid a sum per 1,000 *Schüsse* (shots of the shuttle across the warp). Culturally this was reflected in the colloquial term 'looking for a position' when seeking work, and the changing out of street clothes into company work clothes further emphasised the employers 'ownership' of the employee while on the factory premises.

Did these different national conceptions of labour have any impact on the development of the construction industry in Germany and Britain? The role of the *Deutscher Werkbund* in the 1920s was influential in transmitting the manufacturing ideology of mass-production, particularly standardisation, and sequential assembly into German architectural practice. The German concept of labour, which would include the worker and his actions as part of the technology of production, was therefore far more conducive to the success of rationalising the building process, which is apparent from the early legislation on industry standards (DIN codes) and the work of Ernst Neufert (Chapter 5). Following on from this, if the attributes and skills embodied in a worker are appropriated by the employer and transformed through the employment contract into the product, then the education and training of workers become of critical interest to the employer. This in part would explain the better provision of vocational training and technical education for building workers in Germany.[39]

In Britain, however, the wage of the British construction worker was based on productive output and the worker retained ownership of his labour power. For example, the daily rate paid to a bricklayer was paid on the evidence of what he produced. British construction workers successfully resisted the introduction of piece-rates to construction for decades until they finally were agreed under wartime conditions in the 1940s. Construction employers frequently conceived of labour as a variable input to the construction process, something to be bought when necessary and not intrinsic to the production process itself. In this scenario, education, training and qualifications become almost irrelevant as a worker is employed on the basis of *what* he produces, not through the transformation of his *capacity* to produce while under a contract to an employer.

Sociologically, the notion of the social construction of 'skill' now predominates, implied through the organisation of the social partners, wage structures, collective agreements, employment and contract conditions, different forms of work organisation, and the classifications embodied in educational and training systems (Brockmann *et al.*, 2011). 'Skill', observable in the work process, relates to the complexity of tasks undertaken, and represents the outcome of this process of social negotiation. However, the earlier, mechanistic, understanding of skill was prevalent in the post-war era of industrialised building and is still prevalent in today's construction industry.

Previous accounts of the post-war building programme and the huge changes in industrial techniques it incorporated have not positioned the social structures and processes integral to these changes in the foreground. This book seeks to show that building site culture, and the structure of the industry, stem from productive practices originating in an understanding of labour valued through output. Hence the necessity of the quantity surveyor to measure output, the payment of bricklayers by the number of bricks laid per day, the payment of labourers' by the length of trench dug, the payment of joiners by the number of doors hung in a day, and so on. This conception of labour value by-passes the necessity for education and training, apart from short courses in different techniques, and facilitates a tendency to rely on a casual workforce as the worker him/herself does not embody value to the employer – only the output they produce. Workers and employers trapped in this understanding will argue over wages and bonuses, not rights to education and training. There were, of course, examples of traditional firms with stable, directly employed workforces, apprenticeship schemes and internal career pathways. In not relying on casual labour as the main workforce, these firms provided a different model in terms of employment and training, and they were often responsible for restoration and new-build schemes requiring high levels of skill, but by the 1960s they also measured output in the same way and supported entire departments devoted to productivity calculations. So while the above can explain many of the characteristics of the mid-twentieth-century construction industry in terms of employer/employee relationships, where does the architect stand? This book argues that the architectural profession was an associate in the maintenance of this framework through its position in the social relations of production: a position it sought to maintain as the controller of the construction process while becoming increasingly indifferent to site processes; and in the face of industrialisation, seeking to sidestep site processes in favour of factory manufacture. The entire social structure of the construction process was held in stasis by the rigid, hierarchical class divisions of British society.

The following chapters develop these theoretical understandings of innovation, skill, work, and the changing relationship between architect and operative in the specific context of modular, light and dry, industrialised building in the mid-twentieth-century British construction industry.

Part I
Industrialisation and the British building industry

1 The industrialisation of building

> Building is only organisation:
> social, technical, economic, psychological
> organisation.[1]

The term 'industrialisation' applied to building implies the transition from a primitive or simple era of production to one that is more advanced and usually understood as involving a rational approach to the processes required to produce buildings more effectively and more efficiently, with the expectation that they will also be cheaper.

In Britain, the term came into common usage in the 1950s but it was notoriously difficult to define and commentators struggled to include all the relevant factors. Roger Walters, an early proponent of industrialisation in the architectural press, proposed: industrialised manufacture + industrialised assembly = industrialised building.[2] This formula included the entire building process from raw material to site assembly and Walters interpreted the term 'industrialised assembly' as having two potential meanings. First, the use of machinery on site to increase efficiency but with many of the final operations, for example, the placing of concrete and the laying of bricks, still being done by hand, and second, in a 'hypothetical future', the simple assembly of machine-made, fully finished components. Site mechanisation was very low throughout the 1940s and early 1950s despite the promotion of innovations, for example, mechanised wheelbarrows and brick hoists, in the architectural and building press and Ministry of Works' advisory booklets[3] (Figure 1.1). It was the 'hypothetical future' of interchangeable components suitable for a wide range of building types that inspired many architects in the post-war period. Walters was typical of many in assuming that modular co-ordination of these components was a prerequisite of industrialised building.

This approach to transforming the process of building was, of course, not new, and examples of rationalisation through the prefabrication of components or parts of buildings can be found throughout the history of construction.[4] It is questionable exactly when the introduction of prefabrication into the building process occurred. Some commentators include the pre-cutting

Figure 1.1 Cover of *New Ways of Building* (1948), edited by Eric de Maré, London: Architectural Press

and assembly of timber frames in fourteenth-century house building. The building of Georgian London included the prefabrication of sash windows and other timber components (Clarke, 1992; McKellar, 1999) and the mass-production of stone ornamentation in Eleanor Coade's factories (Summerson,

1991). By the end of the eighteenth century, many of the characteristics, both technical and contractual, of a rapidly modernising industry were emerging (Satoh, 1995). Early nineteenth-century prefabrication was typified by the export of timber 'colonial cottages' to Australia, South Africa and other outposts of the British Empire (Herbert, 1978). By the 1840s and 1850s, the introduction of cast iron as a building material saw the increasing use of factory-made components, and in England the building of the Crystal Palace in 1851 was paradigmatic of prefabrication at component level. Here, the myriad small cast iron components that made up this extremely complex structure were prefabricated in an industrialised process, that is, they were manufactured under factory conditions by steam-driven machinery, and the rationalisation of the building process extended from the factory to the organisation of their assembly on site (Peters 1996).

In Britain, the years between 1914 and 1942 were a particularly fertile period for experimentation in prefabrication and standardisation. Patrick Abercrombie, Stanley Adshead and Charles Reilly, all based at the Liverpool University School of Architecture, promoted standardisation of house building in the architectural press (Swenarton 1989: 33–40). *The Tudor Walters Report*, published in 1918 under the chairmanship of Raymond Unwin, is generally known for its recommendations on space standards and design but it also contains extensive detail on materials and methods of construction. One of its most far-sighted recommendations was that efficiency would be increased through better site organisation, accurate costing and regular employment for building trade workers.[5] The following year, the Local Government Board appointed the Committee for Standardisation and New Methods of Construction. This committee investigated and reported on a large number of innovative building components and housing types that were revisited in the 1940s in the Ministry of Works' *Post War Building Studies*. Research on standardisation was continued in 1924 under the Committee on New Methods of House Construction, set up by the Ministry of Health, which published four reports in 1924–1925 on the relative merits of housing in steel (including the Weir, Atholl, Thorncliffe and Telford houses), timber and concrete and a wide range of pre-cast concrete blocks and slabs. It is notable that the members of this committee included two representatives of organised labour, Richard Coppock (NFBTO) and George Hicks (AUBTW), neither raising objections to the new materials and methods recommended in the reports.[6]

The ultimate success of factory-based mass-production was achieved in the US automobile industry in the early years of the twentieth century. There were parallel developments in Europe where the relationship between design theory and production process was central to the development of the Modern Movement. The principles of mass-production promoted by Henry Ford, together with F.W. Taylor's theories of scientific management, were well known, but the role of the *Deutscher Werkbund* was equally influential in the formation of a specifically German rationalisation movement. Schwartz

(1996) has documented the cultural discourses on mass culture and the struggle for ideological control between the members of the pre-1914 *Werkbund* and, by the early 1920s, its resolution so that the mass-produced type became the goal of all design activity.[7] The 'type' was the key to the architectural interpretation of mass-production.

> *Typisierung* means agreeing on certain forms for doors, windows, hardware, room sizes, installations, plans and finally entire buildings; norms means quantifying these types. [Types and norms] are a self-evident result of machine production.[8]

The architectural manifestations of this were evident throughout Germany in the 1920s, for example, in the *Siedlung* built by Gropius in Dessau and the work of Martin Wagner in Berlin. With the exodus of European intellectuals from the Third Reich in the 1930s, America became the home of many Modern Movement architects. Here Gropius continued his experiments in industrialised production of housing with another German emigré, Konrad Wachsmann, but these failed to become commercially successful (Herbert, 1984). Wachsmann, however, continued his investigations into industrialised building and design, set out in great detail in *The Turning Point of Building* (1961).

Henry Ford's production methods influenced the ideas of the American industrialist and writer, Alfred Farwell Bemis, whose work became known in British architectural circles mainly through the publication of his trilogy *The Evolving House*.[9] Volume III, *Rational House Building* (1936), became a key text for architects committed to standardisation and prefabrication, and is notable for being the first proposal arguing that industrialisation required a standardised module as the basis for progress. Bemis suggested the size of the module as either 4″, derived from the size of a timber stud, or 3″ as derived from a brick plus mortar and it was this proposal, in particular, that made a resounding impact on British architectural thought on industrialisation.

Less well known, but equally important, was Alfred Bossom's account of skyscraper building published in 1934, where an early manifesto for industrialisation was set out, which, as well as standardisation and prefabrication, argued for efficient site organisation.[10] Based on his experience in the USA, Bossom recommended that architect and contractor co-operate from the outset of a contract, with detailed drawings prepared in advance of site work together with 'Time and Progress Schedules' and long-term agreements with operatives' organisations and materials suppliers. Regarding building as of necessity a co-operative enterprise that was best entered into as teamwork, he understood that in Britain, with its rigid class structure, this ideal was going to be difficult to achieve: 'A hundred reasons of social classification and professional pride have made this ideal of co-partnership, this sense of common interest, difficult to attain and to act upon in England.'[11]

During the Second World War, the relatively recent experience of the difficulties encountered in reconstruction after the First World War led to a number of government committees being set up as early as 1942. Investigations into solutions for the coming renewal of the built environment began with the knowledge that the condition of the construction industry at the end of the war was going to be almost identical to that of 1918. The most important of these committees, in terms of prefabrication and standardisation in building, were the Directorate of Post-War Building and the Directorate of Building Materials, both under the aegis of the Minister of Works. Together, these two bodies undertook a large programme of research into innovative building methods culminating in the publication, between 1944 and 1946, of 33 volumes in the series *Post-War Building Studies*. Much of the experimental work published in *Building Studies* was undertaken by the BRS, and as Saint (1987) points out, the Garston Headquarters of BRS between 1943 and 1945 were to serve as a form of informal finishing school for a number of socially and technically minded architects who later held influential posts in the rebuilding of Britain.[12]

Planning for reconstruction

Towards the end of the Second World War, a plethora of handbooks and guidebooks aimed at educating and informing the British people on the shape of the new post-war urban and rural environment were produced.[13] The government's White Paper on Housing, published in 1945, laid out a post-war reconstruction programme of 750,000 dwellings to provide a dwelling 'for every family desiring one'.[14] A further 500,000 new dwellings were needed to complete the slum clearance programme, based on the assumption that housing output would rise to a maximum of 300,000 dwellings by the end of the second year after the war. Meanwhile, the immediate crisis of homelessness in the aftermath of war damage was to be alleviated by the erection of 158,748 temporary houses produced by the Ministry of Works and the Ministry of Supply[15] and a massive programme of repairs.

The production of this enormous output of permanent new dwellings, at a time of massive shortages of both skilled labour and materials, was to be brought about through the use of innovative methods of construction. These included various techniques of prefabrication in concrete, timber or steel and alternatives to 'traditional' construction, for example, *in situ* concrete. The standardisation of plans, specifications, components and the production of codes of practice were also proposed as means to increase efficiency. The shortage of skilled building labour was to be solved by increased recruitment to apprenticeship schemes and the retraining of adults on short-term construction courses to boost the construction labour force to the 1.25 million considered necessary to complete the reconstruction plan.[16] These changes were anticipated as necessary precursors to the modernisation of an industry perceived by the government as having low productivity or, in the terminology of the late 1940s, functioning well below 'productive efficiency'.[17]

The Labour government of 1945–1951 had difficulty in meeting the housing demands of reconstruction, and housing output did not start to rise until the Conservatives came to power. This has been blamed on shortages of skilled labour, materials and the licensing system for new building works but also on a lack of centralised control of the industry so that building workers employed in small firms spent more time on repairs to bomb damage than on new-build (Rosenberg, 1960). However, the repair and refurbishment of houses were a remarkably efficient operation and very successful in providing immediate shelter to a large number of people, particularly in London (Bullock, 2005). Conservative policy did not initially promote industrial methods, however, but relied on traditional brick construction with higher output achieved by a reduction in (social) housing standards, particularly their size, epitomised by the introduction of 'The People's House'.[18]

The role of architects in reconstruction was to be significant both at government policy level and as leaders of local authority programmes. In both positions they acted as servants of the state rather than as independent professionals. A new generation of architects emerged, trained in the late 1930s and 1940s, and keen to embrace whatever changes were necessary to rebuild Britain. Roger Walters was one of them: he had become interested in the ideas of industrialising the building industry after reading Bemis, while he was studying at Liverpool under Reilly, but was not allowed to write his thesis on the subject. His career in architecture spanned the early experiments in 'light and dry' school construction, the NENK system devised by the Ministry of Public Building and Works and the mass housing developments of the 1970s when he became Chief Architect to the Greater London Council and subsequently gained a knighthood. Looking back on the post-war years, he identified a specific set of economic, political and social factors that contributed to industrialisation.

> It was a period of economic expansion, high demand on the building industry, shortage of skilled labour, rising population, especially high demand for public housing and schools in the aftermath of war and there was a need for slum clearance. Added to that, the Labour government had ensured that all county offices should have a good architect's department and the whole attitude to public sector architecture had changed. It was now regarded as respectable. In fact, many architects took the view that architects should work in public service. I was one. So there was a very powerful motivation to make some sort of contribution to post-war society in the minds of many architects.
>
> But the point I want to make is that the combination of very large public sector programmes under the control of central and local government combined with the other issues is not likely to happen again.[19]

Industrialisation continued to be promoted to the construction industry when the Labour Party returned in 1964 under Harold Wilson. Contractor-led, pre-cast concrete systems building rapidly became the public face of industrialisation (Figure 1.2) and the first edition of a journal epitomising this era of British building, *Industrialised Building Systems and Components*, contained the following definition by the contractor Peter Trench:

> Industrialisation as a process must cover everything from the use of ready-mixed concrete through modular co-ordination to the closed proprietary building system. It is, in fact, too wide a term to be of real use but since we have no other to describe the change in the nature of building which is going on we must accept it. It describes in fact the change of an industry based on craft concepts to one based on scientific engineering principles – it is the ultimate philosophy of parts instead of pieces.[20]

The idea of 'process' resonated with the modern image of factory production, conveyor belts, flow charts, multiple components, and fast, clean erection of lightweight buildings. 'Process', for many architects, was an important and compelling part of design and contributed to a unique aesthetic. Less glamorous, but equally innovative, was the introduction of more efficient forms of site and work organisation in the post-war period. Central to these was the idea of operational research developed in the field during the Second World War and, transferred to civilian life, defined as 'a scientific method for providing executives with a quantitative basis for decisions'.[21] The BRS, in conjunction with a number of large firms providing the means for fieldwork studies, pursued this as a means of increasing productivity and efficiency.[22] It was particularly successful in programming work around mechanisation and resulted in the use of critical path diagrams for detailing the sequence of activities in the building process. The use of this method helped to highlight the critical interactions at the start of any building project between design and production and became a tool used by progressive practitioners to argue for greater integration between members of the building team. The graphical method used presented the flow of work as a system, ideally as a single interlocking process through which labour and materials flowed as a continuous operation. However, this was rarely achieved due to blockages and hold-ups caused by late materials, lack of appropriate labour, poor weather, and all the usual exigencies that combine to make building very unlike a factory process. The reasons for this have been summed up succinctly by Alan Tuckman, describing building as a process unlike mass-production which 'with a fixed base for production and a continuous flow of the product, site-based construction has a transience necessitating a variation in tasks and a discontinuous flow of machinery, workers and materials' (Tuckman, 1982).

Nevertheless, by 1965, the Ministry of Housing and Local Government was using a comparison with manufacturing to define industrialisation, 'the

24 *The industrialisation of building*

Figure 1.2 Trade advertisement for system building, early 1960s

term industrialisation . . . covers all measures needed to enable the industry to work more like a factory industry'.[23] This was part of an MHLG circular in support of the Labour government's decision to launch a drive to increase and improve industrialised house building in the public sector.[24] The 'measures' included the use of new materials and techniques, the manufacture of large factory-produced components, dry processes and increased mechanisation. They also referred to the need for improved management techniques, the correlation of design and production and better site organisation, all matters that had been aired many times over the preceding decade, but here, for the first time the operative workforce were included in a definition of industrialisation: 'Not least, industrialised building entails training teams to work in an organised fashion on long runs of repetitive work, whether the men are using new skills or old.'[25] This implies a complete reorganisation of building sites from their typical muddy chaos into streamlined, efficient, factory-like Taylorist sites of production, irrespective of the level of skills used. Judging by the annual statistics on output of industrialised housing systems, the Wilson government's drive to modernise the construction industry was very successful. Table 1.1 shows that in 1968, 41.3 per cent of all new housing starts were in industrialised systems, and in 1970, 41.3 per cent of all completions were industrialised. After this, with a change of government, support for large-scale building of public housing was withdrawn and there was a steep decline in industrialised housing.

The MHLG included 'rationalised traditional' as an industrialised system but 'rat-trad', as it was popularly known, consisted of traditional brick and block construction that demonstrated increased efficiency through rationalisation of the building process using mechanisation and better organisation. Throughout this period of industrialisation, often seen as the era of pre-cast concrete, high-rise building, the system that produced the largest number of dwellings was Wimpey's No-fines. This was not a component-based system

Table 1.1 Industrialised dwellings: local authorities and new towns in England and Wales

	Housing starts		Completions	
	Number	% of all dwellings	Number	% of all dwellings
1964	27,899	19.9	17,171	14.4
1965	37,214	26.6	25,527	19.2
1966	49,406	33.1	37,494	26.3
1967	65,892	39.4	49,049	30.8
1968	61,369	41.3	50,569	34.2
1969	53,666	40.0	53,150	38.0
1970	28,796	25.1	55,701	41.3

Source: Department of the Environment, Housing and Construction Statistics 1970–71.

at all but rather a way of building using *in situ* concrete, introduced from the Netherlands in the 1920s, and used very effectively by both Laing and Wimpey in the 1930s and throughout the 1960s.

Architects and industrialisation

For architects, the prospect of rapid industrialisation brought a specific set of worries as to their position as overseers of the building process as suggested in this 1945 lecture at the RIBA:

> [W]e cannot produce a good entity unless we have one controlling brain, and I think all of us still consider that that one controlling brain should be the brain of a man with an artistic outlook and a sensitive feeling for the needs of the community, for the right use of materials and for the beauty of the countryside or the grandeur of the city, and we still believe that that man is the architect.[26]

This pervasive idea that there should be one person in charge of the building process continued throughout the period studied here and contributed to the difficulties that architects encountered in defining and defending their professional authority and sphere of influence. On an individual level, it contributed to a sense of failure at the inevitable inability to live up to the ideal of the all-rounded Renaissance Man. In the words of Misha Black, 'a further cause of architects' anguish is the concept of the architect as the complete man – the genius who is the expert of all building crafts, planning, etc. . . . Most architects are bred to a more normal pattern.'[27] In Black's opinion, the solution to this anguish was for architects to specialise, in the manner of scientists and engineers, in the different aspects of industrialised building. This solution, however, did not penetrate into the educational framework of an architect's training.

The architectural debate on industrialisation was framed by two opposing viewpoints. On one side were those architects who considered industrialisation (though not dismissing the need for modernisation in terms of new materials and techniques) as an unjustifiable intrusion into the freedom of the designer, and on the other, were those who accepted the pressing need for the industrialisation of building and wanted to ensure that architects played a central role in the inevitable changes ahead. Neither standpoint queried the proposition that the architect should control and lead not only the design, but also the whole building process.

Those in favour of industrialisation argued that the architect should be involved in order to retain control of the aesthetics of building, traditionally the function that no other construction professional was qualified to undertake. In order to retain this position, architects should be in the forefront of deciding on the nature of technical changes, particularly the dimensions of any module and associated components that would be introduced to industry. Ideally this necessitated consensus from both designers and manufacturers on a range of standardised sizes that would then be ratified through legislation, but debate was initially hampered by a lack of consensus on basic terminology. For example, there was misunderstanding between architects of the nuances implied by such terms as, *planning grid* or *grid-plan*, *dimensional co-ordination* or *modular co-ordination*.[28] A more fundamental gulf became apparent between architects and manufacturers. For architects, dimensional relationships were inextricably linked to human proportion, whereas for manufacturers they were part of an existing set of norms relating to the imperial system of measurements.

In the USA, the American Institute of Architects had taken the lead in developing and promoting modular co-ordination, and in Finland the Architects Institute became the official authority for drawing up all the industry standards for building. In Britain, however, it was the British Standards Institute (BSI), which was responsible for the research and introduction of new standards into all aspects of manufacturing. Within the complex committee structure of the BSI, the Building section started work on modular co-ordination in 1947. The first BSI Report on the subject was published in 1951 and recommended the use of a planning module of 3 feet 4 inches, leaving the question of a module size for components open.

The only other state-funded organisation undertaking research in construction was the BRS, where a working party on modular co-ordination was not set up until 1953. The RIBA, despite organising research committees under the Architectural Science Board, did not engage directly in research and the problem of introducing modular co-ordination, with architects in control of the proceedings, was approached in a particularly British way, through the setting up of a private society: the Modular Society.

The formation of the Modular Society in 1953 provided the first forum in which the issues around standardisation could be debated between all the concerned parties, architects, engineers, manufacturers and contractors. The idea of a private society to investigate these issues was originally suggested by Mark Hartland Thomas at the end of a lecture he had delivered at the RSA, entitled, 'Cheaper Building: The Contribution of Modular Co-ordination', in December 1952.[29] He went on to found the Modular Society, with Lord Alfred Bossom, on 19 January 1953, which acted to campaign and promote the use of modular components for building: a major concern of those committed to industrialising the process of building.

Towards the end of the 1950s, Cleeve Barr reflected on the failure of the 'non-traditional' house building programme instigated at the end of the war to transform the building industry and recorded in the *Post-War Building Studies*. He considered the emphasis of research on individually complete houses, rather than on component parts that could be used for a variety of building types, a major disaster from an architectural point of view. He concluded that what was still needed was 'not factory-made houses but factory-made components . . . on the basis of which productivity can be doubled and, at the same time, a decent urban environment created'.[30]

The following account in Part II focuses on component building, or assembly, rather than 'wet and dry' processes, for example, *in-situ* concrete, and documents the early investigations carried out by the Modular Society into the use of the 4-inch module as a basis for building and designing with industrially produced components. Terms used in modular construction are defined in the Glossary on p. 188. By the mid-1960s dimensional co-ordination (dc), as it was then known, had been assimilated into government-promoted systems, for example, 5M and 12M housing, originating from the Ministry of Building and Public Works. New Ministry of Defence building contracts were in dimensionally co-ordinated systems (NENK), and a series of design guides setting out the principles of dc design for local authority housing were published. The architect-led ideology of 'light and dry' component building, first used on a large scale in the post-war Hertfordshire school building programme, had permeated from local into central government strategy and policies.

The building unions and industrialisation

The majority of the building craft unions formed the NFBTO in 1918.[31] Ferdinand Zweig, in 1951, described unions as 'repositories of collective memories' and also acknowledged the position of the Permanent General Secretary as a powerful force with the potential for autocratic leadership.[32] This was certainly the case in the history of the NFBTO where, from the early 1920s until he retired in 1966, Richard Coppock held the post of General Secretary. Wood (1979) describes how, at NFBTO annual conferences, the only people on the platform would be Coppock, his personal secretary and his research officer, Harry Heumann. Famous for his ability as an orator who rarely used notes, he was the construction industry's chief negotiator on the National Joint Council for the Building Industry (NJCBI) for nearly fifty years. However, he was a traditionalist in his views and, through his position as General Secretary, maintained the *status quo* of a federation of craft unions rather than supporting the birth of an amalgamated industrial union. He was politically astute in not condemning industrialisation. At the 1950 annual conference, he refused to support a motion demanding the condemnation of non-traditional building, saying:

> We cannot set ourselves against the forward line of development, but we must maintain as a craftsman's organisation, the basis of our form and organisation, for the purpose of producing, if I may say so, the city beautiful.[33]

This statement, referring back to ideas current at the beginning of the century on the value of craftsmanship in producing beauty, also contains the kernel of the problem that was to slow the pace of modernisation in the construction unions. Coppock believed passionately in the superiority of the skilled craftsman over the labourer, and refused to contemplate the merger of the civil engineering labourers' unions with the craft unions, which dominated the NFBTO. This resistance maintained a fragmented union structure and exacerbated the problems over wage differentials for, as early as 1950, labourers working on concrete and driving plant were earning far more on 'plus rates' than craft rates of pay.

After the high levels of consultation experienced during the Second World War, the building unions proposed, and expected, nationalisation, however, this did not materialise (see Chapter 2). The NFBTO objected strongly to the housing reforms of the Conservative government of 1951, especially the introduction of 'The People's House'. The lowered standards and dimensions of social housing were resented and the 1952 NFBTO annual conference declared:

> This conference desires to make it clear to the Government and the public that building workers cannot be willing participants in such anti-working class legislation, and it declares that the only final solution to the nation's housing problem is the transformation of the building industry into a great social service with its products being allocated on the basis of need.[34]

For the unions, the onset of industrialisation was integrally bound up with the issues of skill, amalgamation and training. The question of industrial unionism cropped up annually at every NFBTO conference, usually proposed by one of the general labourers' unions. The debate invariably included speakers from the craft unions opposed to the motion arguing, first, that their members worked in a range of industries not just construction, and, second, that wages in any industry-wide union should be based on craft, not labourers', rates. Nevertheless the motion for a united union was usually passed if not acted upon. Craft unions defined their membership on the basis of materials worked and with the increasing use of new materials and processes, this inflexibility caused problems.

The fundamental problem for the craft unions revolved around an outdated concept of apprenticeship training that continued to define the skill of the craftsman in a manner very similar to the romantic image of Ruskin's craft workers. As late as 1962, Harry Weaver, who was Coppock's successor at the NFBTO, was defining a craftsman as 'a worker with the manual skill and dexterity required and exercised for the creation of a finished product from its primary sources of raw materials, combined with knowledge and appreciation of such a creative gift'.[35] This implies not just a vocational, but a spiritual element to the craftsman's work, and while it did not reflect the reality of the working conditions on many industrialised sites, it did reveal Weaver's traditional views. He believed that athletic skill was the best basis for becoming a good craftsman and, as in any sport, the younger the apprentice, the better, and 14 being the ideal age.[36] Although the issue of apprenticeship was frequently raised at NFBTO conferences, the building unions did not form any independent, special committees on education and training and, as will be shown later in Chapter 3, the matter of training did not reach the negotiating table in the post-war years.

The post-war construction industry comprised both the civil engineering and the building industries. Distinguished by their different products, the 'civils' produced structures such as dams, motorways and bridges and the building industry produced mainly roofed structures. They also differ in their composition: the civil engineering industry being composed of mainly large firms, and sometimes by their organisation, with the engineer replacing the function of the architect on civil engineering contracts. Both industries had different industrial negotiating bodies and generally operated under different working agreements. At the level of site processes, however, the two, each with their separate wage agreements, overlapped. For example, carpenters and joiners traditionally worked with wood but with the introduction of metal partitioning as an alternative to stud and plaster partitioning, the carpenters union, the Amalgamated Society of Woodworkers (ASW), argued that this process was still within the craftsman's remit and should not be given to labourers. The ASW also argued that the fitting of metal windows was a carpenter's job against the conflicting claims of the plumber's union. Concrete work also gave rise to similar disputes with the carpenters arguing that timber shuttering was the work of skilled carpenters not general labourers.

At the 1950 NFBTO Annual Conference, Richard Stokes, Minister of Works, praised construction industry workers for their acceptance of industrialisation:

> May I pay tribute to all the men in the industry, that, so far as I have observed it, have been more receptive to new ideas and prepared to use mechanical devices, and all the rest of it, with less bother and trouble than any other organised body of labour of which I have had any experience.[37]

This did seem to be the case for most of the unions within the NFBTO, as there were relatively few demarcation disputes that were not resolved at a local level and taken to reconciliation at a national level. The NFBTO organised a conference in 1959 on new techniques in building in which they positioned themselves firmly in favour of progress, as long as this did not mean the exclusion of the skilled craft worker from new processes, and building rates and conditions were recognised.[38] But only three years later there was mounting concern at the rapid increase in the use of heavy pre-cast concrete systems. One delegate to the 1962 conference warned, 'We are undergoing a change that has never been seen before in the history of the building industry.'[39] The conference was warned that the very purpose of industrialisation was to do away with need for any skilled craft workers and there were calls for another special conference on new techniques in order to plan a response.[40]

Throughout the 1950s and 1960s, the debates recorded at the NFBTO conferences reveal that building trades union officials and leaders did not object to the introduction of new industrialised processes *per se* but sought to defend their members' interests in terms of wages and conditions. However, their engagement with the topic was broader than this, and a number of conferences on new techniques were called, delegations to mainland Europe were made to investigate developments overseas, and good workmanship was debated in relation to prefabrication. The building trades unions, together with progressive architects, argued for standardisation of fittings and fixtures in an attempt to cut costs and maintain standards for working-class housing.[41] In 1963, Harry Weaver, General Secretary of the NFBTO, looked forward to industrialisation bringing better planning and continuous production to building, in turn, resulting in an end to casual employment and affirming that 'building workers are not latter-day Luddites'.[42]

2 The building industry during war and reconstruction

> This has been an engineer's war. Let us make it a builder's peace.
> E.D. Simon, 1945, p. 43

Building activity during the war was controlled from 1940 onward through the creation of a new Ministry to co-ordinate all construction, the Ministry of Works and Buildings, which replaced the old Office of Works.[1] Although it was proposed that this new Ministry would control all building work, there was great reluctance on the part of the Minister of Health (Henry Willink, Conservative Minister in the coalition government) to relinquish any control of housing. Housing remained the responsibility of the Ministry of Health until the formation of the Ministry of Housing and Local Government (MHLG) in 1951.[2]

It quickly became apparent that, once a priorities system for building works had been agreed, the Ministry of Works' ability to co-ordinate building rested on knowledge of the availability of labour. The existing statistics held by the Ministry of Labour were very poor and gave no indication of the nature of employment or the whereabouts of the workforce and so, under Lord Reith, the Ministry of Works set up its own statistical department. Labour allocation for future work was based on these statistics, gathered monthly, and strategic planning by the Ministry of Works was turned into operational reality through the Ministry of Labour via the local employment exchanges. The Ministry of Labour had, by 1940, wide powers to control employment, and building workers had to seek work through employment exchanges and register when unemployed. Employers, in turn, were forbidden to advertise and could only engage workers through the local employment exchange.

Wages and conditions of employment continued to be dealt with through the National Joint Council for the Building Industry. Although the term 'building industry' usually includes both civil engineering and building, these two parts operated quite separately, having their own wage bodies, consisting of employer and employee representatives, and different terms and conditions of employment. It was known that both building and civil engineering labour was going to be needed on the defence contracts and potential industrial disputes were minimised by a uniformity agreement. This meant that wages, hours and working conditions were uniform for all workers on government contracts for the duration of the war.[3]

Communication on general matters was difficult to organise for an industry consisting of large numbers of small firms, many of them only two- or three-person outfits. The Ministry initially met with industry through the Central Council for Works and Buildings consisting of invited representatives of employers' and operatives' federations, with Hugh Beaver as chairman. Prior to this, the industry had set up its own elected representative body in 1931, the Building Industry National Council (BINC), a more inclusive body composed of architects, builders, operatives, manufacturers and others. Although it continued to act in an advisory capacity to the Ministry, it was not seen as a representative body as it did not include civil engineering. In 1942, Lord Portal instituted a new body, the Advisory Committee of the Building and Civil Engineering Industries, with elected representatives from wider sections of the industry, including the RIBA. In 1945, under Duncan Sandys, the Central Council for Works and Buildings was dissolved and the National Consultative Council of the Building and Civil Engineering Industries set up. The committee and sub-committee structure existing under these bodies was extensive and complicated and consisted at its peak of over 30 bodies in all, advising and investigating in areas including training, building materials, payment by results, building materials and codes of practice. A large amount of this work was focused on anticipating and planning for reconstruction.

The passing of the Essential Work (Building and Civil Engineering) Order (EWO) in 1941 gave the Ministry of Works statutory powers to implement a system of controls. It also introduced payment by results (PBR), an issue that had been vigorously contested by the employees, led by Luke Fawcett of the AUBTW and the NFBTO, who argued that piece work undermined craft skills.[4] The coalition government, nevertheless, forced through the legislation without the backing of the unions. As a conciliatory gesture towards the building unions, it included the provisos that PBR would only operate during the war unless continued by joint agreement, and that no worker would receive less under PBR than under existing agreements or under the guaranteed pay provision of the EWO. PBR was accompanied with the introduction of a guaranteed wage, in itself a major innovation to the construction industry, which had traditionally relied on casual labour.

The overall structure of the industry was seen as especially problematic for organisational purposes during the war, especially the preponderance of one-person firms. It was also argued that this was the reason that mechanisation and efficiency in the industry were low, nevertheless, this structure remained stable throughout the 1950s (see Table 2.1) and well into the 1960s. In 1965, 73.1 per cent of firms employed only 1–10 operatives,

Table 2.1 Structure of the construction industry by size of firm, 1935–1953

Size of firm	1935 Total no. of firms	1935 % of total firms	1935 No. of employees (thousands)	Size of firm	1946 Total no. of firms	1946 % of total firms	1946 No. of employees (thousands)	1950 Total no. of firms	1950 % of total firms	1950 No. of employees (thousands)	1953 Total no. of firms	1953 % of total firms	1953 No. of employees (thousands)
				1-man firms	48190	45.2							
Up to 10	67450	88	255	1–5	34688	32.4	84.5	45524	38.6		37793	35.9	
				6–19	15733	14.7	159.7	45018	38.1	106.4	42046	40	98.7
11–99	7716	10	252.9	20–99	7099	6.6	274.2	18437	15.6	186.0	16816	16	171.3
								7638	6.4	300.9	7102	6.7	277.1
100–499	868	1.2	163	100–499	1059	0.9	196.9	1065	0.9	200.9	1099	1.0	206.6
500 + over	78	0.1	86.3	500 + over	98	0.1	111.2	133	0.1	179.0	146	0.1	215.5

Sources: 1935 Census of Production; 1946–53, Ministry of Works statistics.

24.2 per cent employed 11–99, 2.2 per cent employed 100–499 and 0.3 per cent employed over 500 operatives.[5]

The immediate post-war years saw the emergence of a small number of 'giant' firms, e.g. Wimpeys, Wates, and Laings, with very large operative workforces of over 1,000 employees, although the relative number of very large firms remained low, at 0.1 per cent in 1953 of all firms.

Labourers and women in the wartime construction industry

The building labour force decreased from 1.206 million workers in 1939 to just 632,000 in mid-1945 (*Labour Gazette*) or 535,000, according to Ministry of Works statistics derived from the industry census. Discrepancies between these two sets of statistics are frequent and probably arose due to the failure of firms to return the census forms while the Ministry of Works statistics made no allowance for non-response.

By 1941, the industry was short of 50,000 building labourers who were not initially included in the schedule of reserved occupations. This shortage became acute with the entrance of the USA into the war at the end of 1941 when a vast amount of work became necessary to provide accommodation for American troops and build the airfields and other defence works needed in preparation for the invasion of Europe.

Mechanisation was low in an industry that had depended on the brute strength of the labourer for practices almost unchanged for centuries. Correspondingly, the proportion of unskilled workers in the pre-war industry was high, at 50.5 per cent in 1925 and 48.9 per cent in 1938.[6] This had declined to 44 per cent at the end of the war in 1946, and then started to rise in the early 1950s to 46.8 per cent in 1953.[7] Labourers, classified by the Ministry of Works as 'unskilled' or 'other' occupations, were employed largely in the building and civil engineering firms. In 1949, the small number of firms employing over 500 operatives classified 73.6 per cent of their workforce as labourers, and this pattern of employment did not change until the recession of the 1980s. Industry statistics chart the apparent decrease of skilled workers in the industry from 1964 onwards, resulting in a greater proportion of unskilled and semi-skilled. While this change occurs over the same period that saw the introduction of heavy, pre-cast concrete panel construction for housing and a corresponding increase in the semi-skilled civil engineering occupations, it should be treated with some caution: this was also the period when very large numbers of skilled construction workers 'disappeared' from the official records into the bogus self-employment (known as the 'lump') that became prevalent in the mid-1960s.

But alongside the 'brute force' interpretation of the labourer's role in construction, there is another reason for the surprisingly high proportion of labourers to skilled craftsmen. All those occupations that did not have an apprenticeship route to skilled status remained classed as unskilled, and this included all the new occupations associated with innovations in materials and techniques, which had been introduced, including concrete, during the early years of the century. There was, however, widespread awareness that this type of work did require skill and the labourer was a central and pivotal figure in all new construction work. The labourer, therefore, held a very important, if somewhat ambiguous role in the building industry. In the opinion of the wartime President of the BINC, Lesley Wallis:

> There is a popular misconception that anyone can be a builder's labourer, that it is a totally unskilled job. The builder's contention that although not skilled to the same extent as a craftsman, the builder's labourer is certainly not unskilled is now realised even by those who in earlier days refused him a reservation status.[8]

The role of the labourer was also significant to those who decried the primitive methods of working prevalent in construction. Lord Simon, trained as a civil engineer, acted as the chairman to all the sub-committees under the Central Council for Works and Buildings, and lamented the general use of manual labour in the following passage:

> The handling of bricks in building modern houses is an extreme example of the impossibility of using modern methods. The bricks are generally taken from the lorry by one labourer and transported to where they are wanted by throwing them from man to man. The bricklayer stands on the scaffolding ... the labourer carries the bricks to him on a hod up a ladder. It is almost incredible to an engineer that methods so wasteful of human effort should be employed today.[9]

The loss of labourers to the armed services early in the war resulted in an 'imbalance' of skills which led to the introduction of the temporary measure of 'designation' of construction workers. This measure in itself did much to clarify the essential contribution of the labourer to construction work.

For a short time, between July 1942 and September 1944, skilled craftsmen, mainly carpenters and bricklayers, were assigned to labouring work and paid the basic rate of a craftsman. This was met with resistance from the Treasury as it flew in the face of all agreements on the transfer of workers from one industry to another based on the principle of 'the pay of the job no more, whatever their remuneration at their own trade'.[10] However, the interests of the building workers were well defended by Richard Coppock, General Secretary of NFBTO, who represented building operatives on numerous Ministry committees. He agreed to the introduction of designation as an emergency measure with the proviso that, 'the flat rate of the craftsman would be ... the rate of the town from which the man comes, or the local rate, whichever is the higher'.[11]

Although wages were protected, the nature of the work to which assigned workers were to be allocated was not clearly defined. The Ministry of Works printed propaganda pamphlets explaining the scheme to skilled craftsmen and stating that they would definitely not have to do 'pick and shovel' work but would assist other craftsmen on general building. At the same time the Ministry was under pressure from the Treasury to try and appease the demands from the RAF for labour to work on the building of new runways. A note addressed to the Air Ministry in 1942 notes that Treasury sanction was being sought to place 'designated craftsmen on civil engineering jobs as labourers, although as we all know designated craftsmen can be comparatively inefficient as navvies'.[12] The scheme also antagonised the employers, with the London Master Builders' Association complaining in 1943 of the 'inadequate output of designated craftsmen put on heavy labouring work on building contracts'.[13] The scheme was withdrawn the following year. A hand-written note, presumably by a civil servant, in the National Archives file sums up the scheme's advantages as having fostered a better spirit between labourer and craftsmen, increased flexibility in labour supply, and increased site efficiency. The drawbacks were the higher costs and claims from the contractors.

During this period there was not just an 'imbalance' of skills but also in the composition of the construction workforce itself which consisted of large numbers of older men together with very young lads, as can be seen in the photograph from the Imperial War Museum archives (Figure 2.1). The second photograph from the same series (Figure 2.2) is open to a wide range of interpretations but appears to be of a visiting Colonel exhorting the workers to greater productivity and being met with a less than enthusiastic response. While the photograph of the woman bricklayer (Figure 2.3) provides evidence of women's contribution to and presence in the industry in much the same way as their presence was recorded in the 1914–18 war.

Over the same period, from 1941 onwards, women began to be recruited into the industry to work on contracts for factories, workers' housing and airfields. Employers began engaging women on a range of work from general labouring on site to joinery in woodworking shops.[14] The craft unions responded at a local level immediately to this perceived threat, an example being the Hendon branch of the ASW, who presented a four-point plan to a number of old-established joinery firms in their area restricting the employment of women.[15] But when these terms, which included the demand that women workers should pay their union levy, but be excluded from any meetings, were published in a letter to the rank-and-file journal *The New Builder's Leader* (*NBL*) in May 1941, they were met with a brisk response from the editor, Frank Jackson, asking why the ASW was discriminating against women and playing into the hands of the employers.

The *NBL*, founded in 1935 by Frank Jackson, a Communist and carpenter and joiner, was the mouthpiece of the general building unions. During the war years it held a particularly strong position as an industry newspaper since *The Operative Builder*, the journal of the NFBTO, had suspended publication between 1932 and 1947 because the Executive had decided it

Figure 2.1 Building workers c.1943 listening to a Colonel returned from the Malta campaign

Source: Imperial War Museum Photograph Archive.

was not reaching its target audience. The *NBL* published a range of articles including those from members of craft unions, but it was clear from editorials and special pieces on the role of labourers in the industry that its aim was the promulgation of 'one big union' for the industry. It also appears

Building industry in war and reconstruction 39

Figure 2.2 Building workers c. 1943 in conversation with a Colonel returned from the Malta campaign
Source: Imperial War Museum Photograph Archive.

Figure 2.3 Woman bricklayer in the Second World War
Source: Imperial War Museum Photograph Archive.

to be the only trade union paper, apart from *Keystone*, the journal of the Association of Building Technicians (ABT), to write encouragingly on the role of women in building. Despite these editorial sympathies towards women, union representatives were often less well disposed.

By September 1941, the terms for a new national agreement to cover the employment of women workers had been debated by the NFBTO General Council, with one of the points stressed by the representatives of the Labourers' Unions being that there should be 'provision for ridding the industry of women after the war'.[16] The Labourers Unions had proposed that all women entering labourers' jobs were to be immediately paid the full male rate and any plus rates and also be eligible for any increases negotiated at national level. These terms were not demanded out of any feminist principles but rather, given the employment of women was regarded as merely a wartime expediency, 'in the belief that this provided the best means for getting the women out of the industry after the war'.[17] By contrast, the Constructional Engineering Union (CEU), who dubbed themselves the 'Iron Fighters', prided themselves on their acceptance of women members on the same basis as men.[18]

In October 1941, the NJCBI, consisting of representatives of employers and employees organisations, agreed the terms on which women would be employed in the industry 'during the period of the war'. These were a set of restrictions on the employment of women, requiring any employer to first consult with the appropriate trade union on whether any men were available, and even after employing women, ensuring that if at a later date men became available, 'the number of women may be correspondingly reduced'.[19]

The Ministry of Labour and National Service noted that these conditions on the engagement of women would probably result in very few women entering building employment.[20] The agreement also specified, that the basic rate of wages for women engaged on craft processes was one shilling and sixpence (1s 6d.) per hour, 20 per cent less than the corresponding male rate. Despite these restrictions and according to the Labour Research Department's figures, in 1939 there were 15,700 women employed in the industry and by 1945 this had risen to 24,200, giving a participation rate of 3.8 per cent of the total construction workforce.[21]

Recognition, training and skill in the post-war years

Union leaders downplayed women's role in the building industry during the Second World War, and Richard Coppock, General Secretary of the NFBTO, in his many post-war paeans to the skill and endurance of building workers during the war fails to mention that women were ever employed in the industry.[22] Towards the end of the war and facing an acute labour shortage, the Minister of Labour demanded that all men up to the age of 60 with any experience of building register for reconstruction work, and refused to recruit any women into the skilled trades, despite their recent war experience.[23] In a 1945 *NBL* article entitled, 'Can we find the labour power?' Frank Jackson criticised the government and argued that as well as government training for men returning from the forces:

> [W]e MUST face up to the fact that there are numerous jobs that can be done on building by women and we shall have to recognise their value and help to tackle this really gigantic task now facing us.[24]

Later in the same year, the new Minister for Labour, George Isaacs, addressed the NFBTO annual conference on the subject of 'Augmentation of Skilled Labour in the Building Industry'. In this speech he reassured delegates that all adults who passed through the government training centres had been suitably selected, the content and length of their training were of a high standard and after 14 months with a suitable employer, they were to be regarded as fully skilled. He urged the industry to welcome these craftsmen and not discriminate against them because of the 'novel method of their entry': the term 'dilutee' was carefully omitted from his address.[25] Isaacs also thanked building workers, especially members of the AUBTW, for their vital war work on aerodromes, hospitals and accommodation needed for the arrival of the American Allies, but without mentioning women at this or any other point of his address.

But women remained in the industry and while they may have disappeared from building sites, many thousands worked in joinery factories where they made everything from sash windows and flush doors to the new Uni-Seco prefabricated houses. Unfortunately, industry employment statistics showing a gender breakdown disappear in 1962 when *The Operative Builder* ceased publishing. For most of the twentieth century, statistics for employment in the building industry, produced by government departments responsible, contained no systematic gender breakdown. Census records do nevertheless indicate general occupational trends and show that some women did succeed in staying in the industry, particularly as painters and decorators.[26]

These wartime episodes illuminate the deep-seated problems within the industry in relation to recognition of skill. Both women and labourers were kept out of the skilled craft unions because in both cases they had not been through an apprenticeship, which was used as a barrier to entry and also to maintain higher wage levels. The designation scheme also revealed the industry's dependence on large numbers of 'unskilled' manual workers for its very existence and that the nature of this work, rather than being 'unskilled' demanded a different set of as yet unrecognised skills.

Wartime contracts for airfields and the building of floating Mulberry harbours used in the Normandy landings allowed large firms, Laings, Wimpey and Bovis, for example, to increase their technical expertise, especially in concrete technology. These schemes had required high levels of mechanisation and organisation and the large firms emerged from the war with new plant in anticipation of the post-war programme of reconstruction.[27]

However, industry investment in skills and training of the workforce did not proceed at the same rate as capital investment in plant and machinery, and the high levels of unskilled in the workforce continued into the 1950s and beyond, throughout a period when the construction industry was rapidly introducing industrial methods. This is comparable with the inter-war period when there was considerable change in construction materials and techniques but the proportion of skilled to unskilled did not change at all.[28] The training, or rather lack of it, and the lack of recognition of the skills of large sections of the workforce in the use of new materials and processes became a central determinant in the development of the industry.

Nationalisation

Proposals for Soviet-style industrialisation and rationalisation of the building industry had already been set out in some detail by J.D. Bernal in *Britain without Capitalists* (Anon, 1936). Bernal contributed a lengthy chapter on building to this collection of essays written anonymously by a group of Communist economists and scientists. After describing the current state of the industry and its products as presenting the 'essential defects of capitalism', to the extent that 'a Hogarth and a Grosz in union would find it hard to compose a caricature equal to the situation as it is', he proposes a complete overhaul of the entire system.[29] Bernal's suggestions covering every aspect of construction from building materials to town planning, included eradicating the existing demarcations between architects and operatives and increasing mechanisation to create a fully modernised and industrialised industry. He proposed the reorganisation of working relations for industrialised building based on collective, non-hierarchical roles:

> the relationship between the architectural staffs and the building operatives on the job leaves much to be desired. In a Britain without capitalists the meeting of the architects with the works committees would presumably be an inseparable part of the work, without any need or desire for a pose of superiority on the part of the architects.[30]

The same themes occur in post-war proposals for nationalisation but without the radical and revolutionary fervour of J.D Bernal's ideas on modernisation of the construction industry.

After the high levels of consultation and planning experienced during the war years, the building unions were convinced that nationalisation of the industry was a viable proposal for post-war reconstruction. There were a number of proposals mooted for the transformation of the building industry into a public service: all suggested the setting up of some form of central, national co-ordinating body for building. G.D.H. Cole, for example, in

1945, rejected comprehensive nationalisation of the industry as unworkable and preferred to leave the existing structure of many small building firms but 'organise them as to co-ordinate their efforts and to bring their practices into better harmony with public needs.'[31] Membership of Cole's proposed National Building Corporation, including employers, operatives, architects and all the other professions, was to be voluntary. The field of operation envisaged was mainly housing renewal and one of the primary functions of the corporation was to provide good quality training for apprentices.

In the same year, Harry Barham, a Scottish building trade unionist, aired his proposals for nationalisation in a short pamphlet:[32]

> The only solution of the many problems involved is to convert the building industry from its present condition of multiple warring elements with a tendency to capitalist monopoly into a public service; to eliminate completely capitalist control and private profit; and to substitute a service organised to produce the buildings the nation requires as part of the national plan.[33]

While Luke Fawcett endorsed Barham's proposal, Richard Coppock, writing in the Introduction, was more circumspect, considering that it was 'not a practicable proposition to plan a semi-socialised industry within a fundamental capitalist framework'.[34] Later, in 1947, Barham put together a detailed programme for the complete nationalisation of building, including proposals for a National Building Council and a Building Worker's Industrial Union.[35]

An alternative plan by the Ministry of Works suggested converting the Special Repairs Service, which had been used during the war for emergency repairs after air-raid attacks, into a National Building Corporation.[36] The function of the Corporation, to be allocated a permanent labour force of 100,000 men, was to supplement building labour where needed for reconstruction. It was to be run by a board composed of six members from both sides of industry, responsible to the Ministry of Works. This corporation, however, never materialised due to the antagonism of construction employers who refused to have anything to do with it, fearing it was the first step towards nationalisation. Whereas Hall (1948) described the shelving of the Building Corporation as due to the Ministry of Works refusing to go ahead without the employers taking part, Richard Coppock, writing 11 years later, gave a more detailed account.[37]

When the Building Corporation was first suggested by the Minister of Works in 1946, Luke Fawcett (AUBTW) was appointed Chairman, the other appointees being Richard Coppock (NFBTO), Frank Wolstencroft (ASW), Major-General Appleyard and Montagu Meyer (the timber merchant) – all of whom had served in the committee structure of the wartime Ministry of Works. The primary aim was to jump-start the housing drive through the authority of the Ministry of Health. However, the Health Minister, Aneurin

Bevan, refused to release the allocated funds after discussions with the building and civil engineering employers' associations who made him aware of their strong hostility to the whole idea of the Building Corporation. The employers' objections were, as Coppock pointed out, 'hardly surprising in view of its philosophy based on a non-profit social service'.[38]

W.S. Hilton (1963) cites this episode as marking a souring of relations between the building unions and the Labour government of 1945–51, and in particular Aneurin Bevan, who at the 1947 Labour Party Conference completely dismissed the idea of the Building Corporation as having any use in solving the housing problem.[39] At the next AUBTW conference, a member asked if 'Bevan would have used the same language on nationalisation to the miners as he used to the building trade workers'.[40] When Bevan sided with the employers in the 1947 dispute over introducing incentive payments to the building industry and consolidating the PBR scheme introduced during the war, it became clear that there was little government support for nationalisation.

Nevertheless, Coppock, together with the NFBTO's research officer Harry Heumann, published his own ideas for the post-war industry in 1947.[41] These entailed a continuance of the private sector but strictly controlled by a National Building Corporation with the power to make decisions on the volume and type of work. This would achieve the elimination of speculative builders, the bugbears of the building unions since the 1930s boom. When a scheme for nationalisation was finally put before the NFBTO annual conference in 1950, it was virtually identical to that proposed earlier by Coppock and Heumann.[42] When criticised for the lack of clarity on the role operatives were to take, the authors replied that operatives would not own the industry, as syndicalism was not Labour Party policy. Despite the proposal's adherence to party policies, nationalisation did not occur, although it regularly appeared on the agenda at NFBTO conferences over the next two decades. In 1956, the NFBTO published a pamphlet setting out the advantages of nationalisation aimed at building workers. *Building as Public Service* argued for public ownership of the construction industry and for building for 'social need, rather than building to satisfy private greed'.[43] This document also criticised the BRS for being partisan in that it was funded by the taxpayer but at the service of private industry, which was unwilling to pay the cost of its own research.[44] When nationalisation again became an issue for public debate in the mid-1960s it was rebuffed with a very effective campaign funded by private sector firms called CABIN (Campaign Against Building Industry Nationalisation).

Architects in wartime

In the early years of the war, architecture was a reserved occupation for those aged 25 years and over, providing they were employed in work that utilised their professional qualifications.[45] In the eyes of the government, this included being employed by the large contractors, which had been given contracts for the extensive defence works. Later, architecture was removed from the reserved occupation schedule completely and architects were recruited into the armed forces, with the RIBA attempting to protect its members by recommending that they were to be treated in the same way as engineers.[46]

Some of those who remained in employment were engaged in assessing bomb damage, and early in the war, in 1940, the Ministry of Works under Lord Reith set up the National Buildings Record in an attempt to safeguard and repair damaged historic buildings after reports of unnecessary demolition after bombing raids.[47] Over 300 architects were appointed singly or in panels throughout Britain to prepare lists of historic buildings in their areas and this information was then transferred to Air Raid Protection (ARP) controllers. Many architects were engaged on designing air-raid shelters and there was vigorous disagreement with the government over the most efficient design. Tecton, after commissioning Ove Arup to provide the calculations, promoted large, circular deep shelters, which could house many thousands of people.[48] These shelters were never built: the government's Anderson and Morrison shelters aimed at family-sized units became the ubiquitous and cheaper option. Shelters for mothers and babies were designed by Erno Goldfinger, and Birkin Haward designed a residential nursery school for the under-fives: both schemes were published in *Architectural Design* in 1940 as part of the AASTA's (Association of Architects, Surveyors and Technical Assistants) study of evacuation provision for urban evacuees in rural areas.[49]

Wartime working practices were critical in forging the ideas that were to dominate architectural approaches to post-war industrialisation (Saint 1987). The men who later rose to prominent positions in government and in the post-war school building programme, worked on massive wartime industrial projects in large American-style teams headed by William Holford for Alexander Gibbs and Partners. They were tasked with working rapidly and efficiently to create factories, housing and hostels for munitions workers using standardised designs and parts. Yet another group of architects, also important in post-war reconstruction, were involved in camouflage design and building inflatable dummy tanks, landing craft and lorries used to confuse German intelligence on the location of British forces. David Medd reflected on this experience as providing the insight that 'the designer was a link in a complete chain, not a detached component' (Saint 1987: 21).

As well as the experience of team-working in large groups, the war also brought architects into close contact with tradesmen. Alan Crocker, later to author a handbook on dimensional co-ordination, remembers that it 'made me realise that tradesman were pretty knowledgeable and that made a considerable difference to my attitude to looking after building contracts after the war'.[50] 'Call-up' also subjected architects to the hierarchies of military command and the use of military terminology and this, coinciding rather neatly with the existing norms of building site control, continued into the immediate post-war years. Alan Meikle, who served in the navy with Henry Swain, remembered that naval attitudes and slogans were often carried over into architects' offices in the early post-war years. Strict hierarchies within the building industry continued during the 1950s and no operative would ever be on first name terms with the architect who was always formally addressed as Mr.[51]

The role that architects were to play in reconstruction was widely acknowledged, but predictions of the numbers needed were not produced in the same detail, during the war years at least, as for the construction labour force. In 1947, the Ministry of Labour produced a paper predicting the future requirements of the profession, which concluded that a 'maximum strength of 13,710 architects' would be needed by 1950.[52] Estimates of future newly qualified architects, including those already training and 'natural wastage', arrived at a figure of 12,790 architects in practice by 1950 so that collaboration with the RIBA to control the numbers entering, rather than promote recruitment, was recommended. This was to be achieved through discussion between the Ministry of Labour and the RIBA on the exact numbers required. To this end, a directive from the Ministry of Labour was circulated to architectural panels, responsible for selecting candidates and awarding grants, to restrict the number of approvals. The Ministry representative on the London panel, E.G. McAlpine, worried that architecture would be an 'over-crowded profession' in the post-war years and that 'many of those who have already got awards will not find it easy to make a living after they are qualified'.[53] The Ministry's procedure was to discourage applicants from a career in architecture:

> The London Panel has told a considerable number of applicants that it is not in their own interests that they should begin the long 5-year course leading to the A.R.I.B.A., and I have personally explained to such candidates that if they amend their application and want to be doctors or teachers or in fact take up any other profession except architecture, there will be no difficulty about an award.[54]

However, it seems that this policy was excessively cautious in that by 1950 there were only 12,484 practising architects (RIBA), over a thousand lower than the Ministry of Labour's 'maximum strength' recommendation.

When the Hankey Report, *The Present and Future Supply and Demand for Persons with Professional Qualifications in Architecture* was finally published in 1949, it contained only two major recommendations. First, that no action was required to increase the present rate of entry into the profession so that any increase in the numbers entering should be discouraged.[55] Second, that both central and local government authorities should use private architectural practices for work in excess of the capacity of their own staff. The report summarised the view of the RIBA that the basis for estimating the future size of the profession was to equate the number of architects to the amount of building labour available. This in turn was based on the figure of 1.25 million workers published in the 1943 White Paper *Training for the Building Industry*. The response in the architectural press was antagonistic. An *AJ* editorial described the report as 'slight, ill-timed and ill-considered' and questioned the accuracy of the information which the RIBA had supplied and on which the Ministry based its assumptions.[56] The 4,000-strong Association of Building Technicians had not been consulted and their response to the report was equally damning; 'There is little in this Report which is not open to serious question.'[57]

G.D.H. Cole's report on training for the construction industry

At the height of the war, in the early 1940s, preparations for peace began under the mantle of a vast number of both statutory and voluntary committees. One of the most ambitious of these was the Nuffield Social Reconstruction Committee, proposed by G.D.H. Cole and funded by the Treasury: it started work in 1941. Addison (1994) identified the work of these committees as representing a 'broad movement of progressive thought' allied to the rational ethic of modernisation together with ideals of social welfare. This applies particularly well to the research undertaken by G.D.H. Cole on construction training. Cole's report, presented to the Education Committee of the Central Council for Works and Buildings in November 1941, was wide-ranging in its recommendations and highly critical of the existing state of training.[58] He noted that current apprenticeship training in England and Wales was deficient in not being standardised so that the content of training varied not just between regions but also between firms. There was also no clear age of entry, no defined time in technical school, and apprentices did not have the freedom to move between firms. He also criticised the inadequate training in new and expanding processes. On this basis he recommended:

> that a form of apprenticeship to the trade as a whole rather than to an individual employer may have much to recommend it . . . and that any satisfactory scheme must include a much larger element of formal technical instruction in schools or classes than is usual in practice at the present time.[59]

He also commented that training for skilled craftsmen was associated exclusively with the traditional crafts and that a wider range of occupations deserved to be recognised as involving 'special types of skill'. With insight and prescience, he anticipated the future requirements of industrialised construction. Cole approached the question of the labourer from a sociological position:

> It should be recognised that neither navvies, nor building labourers, are in truth unskilled workers, and that the relative importance of workers

who do not belong to the traditional crafts is bound to increase as methods of construction change, even if the older groups of tradesmen retain their preponderant importance in cottage building. It is of no less importance to maintain a high standard of quality in the newer occupations than in the old crafts, and to give recognition to skill, even where it is not manifested in a traditional form.[60]

Cole was, in effect, recommending an end to traditional apprenticeship in favour of a form of industrial training. The Education Committee, however, do not appear to have spent much time deliberating his proposals for change. Instead the minutes record prolonged discussions between the employers and trade union representatives on the numbers of men deemed appropriate for training on short-term government training courses, the length of these courses and the status of these trainees.[61]

Over a period of two years the Education Committee debated the issue of adult training at length: which skills were going to be in short supply, the content and nature of short-term training, and status of these newly trained men. The long-term issue of reform of apprenticeship training took a back seat to the urgent problem of providing enough skilled labour for the repair and reconstruction of the built environment at the end of the war.[62]

Cole's report on training, together with his investigation into the demand for labour in the post-war years for the Central Council for Works and Buildings, became the basis for the *Report on Training for the Building Industry* (Simon Committee report), published in 1942.[63] The Simon Committee's report was far more detailed than the White Paper that followed a year later and contained a number of recommendations that were not turned into legislation. It noted the importance of a scientific education to enable the skilled worker to adapt to new working conditions and the importance of a broad general education.[64] It also foresaw, in the near future, a 'trend towards public responsibility for apprenticeship as an institution' which, though vaguely worded, alludes to some form of legislation to regulate the institution of apprenticeship. Finally, there is the suggestion that a formal system of training resulting in a recognised register of skilled workers will enhance the likelihood of professional recognition for skilled workers. Citing the examples of 'nurses, teachers, . . . patent agents, architects, veterinary surgeons to name only a selection', it suggested that registration was the first step towards professional recognition and 'that any distinction between a professional and non-professional skill is itself largely a convention'.[65]

These radical proposals did not appear in the White Paper, *Training for the Building Industry*, which was presented to Parliament the following year in February 1943.[66] This short document proposed a construction workforce of 1.25 million, necessary to undertake a programme of reconstruction stretching over 10–12 years. It recommended an end to casual employment and the training of 200,000 men on intensive government courses funded by the state to alleviate shortages in the years immediately following the war. As far as the education of the future construction workforce was concerned, however, the 1943 Act defined the terms that were to distance apprenticeship from mainstream education: 'apprenticeship training, unlike special adult training, will not be provided and paid for by the State and the various questions which arise in controlling apprenticeship are traditionally settled by the industry itself'.[67]

The White Paper endorsed the view that apprenticeship training was 'the recognised method of training in employment and of entry into the ranks of the skilled workers', and conceded that it needed reviewing and systematising. This was to be through the establishment of the Building Apprenticeship and Training Council (BATC) and this body, consisting of employee and industry representatives, together with members of various government departments, started meeting early in 1943.

The size of the post-war labour force recommended in the White Paper was considerably smaller than that proposed by G.D.H. Cole in his research. He had published a detailed account of his findings and calculations of future demand in an article in *Agenda* where he also made clear his sympathies for building workers, their case for the 'guaranteed week' and all-round better employment conditions.[68] After the publication of the White Paper, Cole wrote a second article in *Agenda* setting out his disagreement with and disappointment over the government's proposals, especially the reduction in the number of skilled workers he had estimated as needed for reconstruction.[69] Here he suggests that the drastic scaling down of the number of new trainees proposed in the White Paper has arisen because the government has based its calculations on assuming a decrease in the numbers of skilled workers needed. Changes in technique, increased prefabrication, for example, were generally understood to reduce the proportion of skilled to unskilled workers (prior to the war it was 50:50), though, as Cole pointed out, the inter-war years had not seen a change in these proportions despite technical changes. He commented scathingly on the proposals for the Apprenticeship Council which he perceives as having very little power of enforcement and concludes that 'it cannot be forgotten that the history of the building industry is strewn with apprenticeship schemes which have looked fairly good on paper, but have in practice seldom been observed'.[70]

Conclusion

The experience of wartime working, in large teams and designing with standardised parts and components, provided many architects with a framework for understanding how these new ways of working might improve the functioning of the building industry. For those committed to social justice it would seem that the benefits would not only affect potential users of the built environment but also the building workers themselves: light and dry construction methods were seen as providing a far better working environment than the wet and muck of traditional sites. Wartime work for building workers had also provided glimpses of a better future. Although PBR was enforced by the government despite opposition by trade unionists, there were benefits to wartime work in terms of lodging allowances, free travel to work and in some cases provision of welfare facilities on site or nearby through a system of mobile canteens. The designation scheme, the employment of large numbers of women, the vast building sites for airfields, camps and military installations and the unifying agreement between the civil engineering and building industry also provided for very different ways of working, compared to the inter-war years. Preparation for post-war reconstruction augured well for building workers, hoping for a better training system, full employment and nationalisation. Although the views of the men themselves are difficult to find, the key difference, for the body of building workers, was that their representatives, in, for example, Luke Fawcett, Richard Coppock and George Hicks, were an integral part of the machinery for planning and operating all civil building works during wartime.

3   Education and training

> You completely destroy the *lingua franca* when you go into industrialization. It opened the door to a different set of skills. Once you move away from the traditional trades, with all their problems, but which can usually be corrected by the subsequent trades, you have a different set of skill requirements. But none of this has been absorbed.
>
> Donald Bishop, former head of Production Research, BRS, interviewed in London, May, 1998

Post-war reforms of architectural training

Perhaps more pressing within the architectural profession than estimates of future manpower was the debate over the structure and content of architectural education that emerged in the post-war period. The RIBA set up a Special Committee on Architectural Education in 1939, which produced a report in 1943 steering a 'middle way' between supporters of Beaux-Arts-type education and modernists wishing to reform training in line with the Bauhaus model. By this time the 'Art or Profession' controversy of the late nineteenth century had transmuted into the 'Art or Science' debate of the mid-twentieth century. Alan Powers suggests that Lethaby's ideas on architectural education, based on the first principles of construction materials and practice, were, by the 1920s, superseded in architectural schools by design and drawing as the main part of the curriculum.[1] The 1943 report was at least clear on where the RIBA positioned itself by stating, 'Even in a scientific age the architect should approach his work as an artist aware of science rather than a scientist aware of art.'[2]

Two further reports, *The Ad Hoc Report on Architectural Education*, 1952, and the 1955 MacMorran Report, made slight changes to the existing system, which at that time consisted of three possible routes to professional status. These were, first, by entering articled pupillage combined with part-time study for three years, after which obtaining salaried employment and continuing with part-time study; second, taking a full-time course in a school of architecture, or third, becoming a paid assistant in an architectural office and studying part-time. Professional status, regardless of the route taken, entailed passing the RIBA set examinations. The MacMorran Report recommended that the RIBA extend recognition to the qualifications taken by part-time students from architectural offices to provide exemption from the RIBA qualifying examinations.[3]

Crinson and Lubbock describe the gradual penetration and domination of the RIBA Education Committee by members who were ardent modernists in the mid-1950s, culminating in the 1958 Oxford Conference.[4] They cite the personal account of William Allen describing Stirrat Johnson-Marshall as one of the prime movers after he had been elected to the RIBA Council in 1953. Early in 1950, soon after Johnson-Marshall had joined the Ministry of Education, the RIBA Council had written to the Minister of Education, George Tomlinson, complaining that 'it was not in the public interest for the Ministry of Education to enter the field of architectural practice'.[5] This referred to Stirrat Johnson-Marshall's plans for a new school at Wokingham for Berkshire County Council. Johnson-Marshall asked the Minister to meet with the RIBA suggesting that such an interview would provide the Minister 'with an excellent opportunity of correcting one or two rather unrealistic views held by a great many architects'.[6] He excused himself from any meeting, and put forward a civil servant as his representative on the grounds that 'this is an occasion when the well-informed layman can deal more effectively with architects than can fellow architects'.[7] According to William Allen, it was this incident in particular that rankled with Stirrat Johnson-Marshall. Allen often dined with Johnson-Marshall, C.H. Aslin and others during the 1950s and discussed the radical changes they believed were necessary to modernise the profession.[8]

The 1958 Oxford Conference, held in April at Magdalen College, consisted of 53 invited guests and speakers, among them, Richard Llewelyn-Davies, Percy Johnson-Marshall, William Allen, Leslie Martin and Robert Matthew, and produced six main recommendations. These were disseminated in an account of the conference written by Leslie Martin for the *RIBAJ* rather than through any publication or report of the proceedings. They were as follows:

1   The present minimum standard for entry in to training was too low and should be raised to a minimum of two 'A' levels.
2   Courses based on Testimonies of Study and RIBA external exams were 'restricting to the development of a full training for the architect . . . and should be progressively abolished'.
3   'Recognised' schools should be in universities or similar institutions.
4   Courses should be either full-time or, on an experimental basis, sandwich in which periods of training in a school alternated with periods of training in an office.
5   That the raised standards of education for the architect will make desirable other forms of training not leading to an architectural qualification but providing an opportunity for transfer if the necessary educational standard is obtained.

6   That postgraduate and research work is essential and should be expanded.[9]

In 1962, Elizabeth Layton was employed by the RIBA as Secretary to the Board of Architectural Education to implement the proposals of the Oxford Conference. Full professional status was still acquired by passing the RIBA examinations and so the professional body continued to maintain control of entry to the profession. Although the Oxford Conference was influential, and it can be assumed that many of the participants favoured a Bauhaus-type education as laid out by Gropius, the curriculum content of the British schools was not specified and remained varied.

One of the consequences of the Oxford Conference recommendations, that entry required two 'A' levels and full-time study should take place at a university-level institution, was that the possibility of joint training for technicians and architects was made impossible. The clear separation of architectural technician training and qualifications, which became Ordinary National Certificate (ONC) or Higher National Certificate (HNC) and included 'design appreciation' but no studio work, from that of architects blocked the old route of working up from the office to professional status. The RIBA had, since 1948, represented their interests in the training of technicians by having input into the curriculum content of both ONCs and HNCs through their membership, together with civil and structural engineers and surveyors, of the Joint Committee governing the qualification. The outcome of the Oxford Conference recommendation was the creation of a two-tier occupation, with architects clearly demarcated from technicians by their different route to professional qualifications. This was obviously discussed and anticipated at the Oxford Conference in that recommendation 5 suggests the possibility of transferring from the architectural technician route to that of becoming an architect.

The RIBA set up a Committee in 1960 specifically to deal with 'the problem of the technician in the architect's office', which finally reported in 1964.[10] This concluded that although the principle of providing a bridge into the senior years of architectural study was agreed, 'the RIBA's present belief is that a bridge provided for entry into architecture will not be crossed often, and that the process will be a difficult one only accessible to carefully chosen students'.[11]

At the same time there were calls to increase architectural students' knowledge of building site practice to better equip them for employment. One of the objectives of the Layton Report was, to 'encourage a wider range of training with other members of the building team, particularly builders' as 'the separation of the architect from the rest of the building industry accounts for many of the architects' difficulties'.[12] This, however, did not materialise, as any firm proposal for joint training within universities or colleges. Instead, under the scheme of training suggested by Elizabeth Layton, building site experience was to be a component of the years spent out of university gaining practical experience and not integrated into the five years spent in full-time education.[13]

One of the effects of the Oxford Conference's emphasis on university-level education was to exclude architecture students from any contact with building trade trainees. This was the final nail in the coffin for putting into practice Lethabite ideas on craft and construction as a basis for architectural education. The influence of Lethaby, however, did not disappear altogether and there remained a few schools where, well into the 1960s, an introduction to the building trades was part of the curriculum, Newcastle and Birmingham being examples.

The idea of initial joint training for building professionals was, however, being followed up in some quarters. This was, of course, by no means a novel idea; the most recent airing had been in a conference convened by the National Joint Consultative Committee (NJCC) of Architects, Quantity Surveyors and Builders at the RIBA in February 1956.[14] A few years later, after the publication of the Emmerson Report, and at the same time as Elizabeth Layton was producing *The Practical Training of Architects*, an informal meeting between representatives of the RIBA, the RICS, the Institute of Structural Engineers and the Institute of Builders took place. The subject was a proposal to investigate the feasibility of joint training in the building industry. The NJCC of Architects, Quantity Surveyors and Builders, duly agreed the setting up of the Joint Committee on Training in the Building Industry in May 1961. Chaired by Sir Noel Hall and fielding William Allen, Professor D. Harper and Edward Mills from the RIBA, and including Elizabeth Layton acting as RIBA Under-Secretary for Education, as well as representatives from the other three professional bodies, the Committee was charged with three terms of reference. These were:

1 Courses which, with suitably varied options, may produce a man of common usefulness, or nearly so, to the fields of both architecture and building.
2 The seconding of trainees in architecture, building, engineering and quantity surveying from their own discipline to that of another for a period.
3 The possible development of joint education/training in a more thoroughgoing way at, broadly, university undergraduate level.[15]

The investigations centred on (3) and involved collating and comparing curricula from all the institutions training students for eventual membership of the four professional bodies involved. The Committee presented their findings in 1965, the main recommendation being that, 'regardless of differences in detail, there is a strong common subject content running through the syllabuses of the four institutions and that teaching and examining should express this fact'.[16] The rest of the recommendations supported this by asserting the importance of implementing common syllabuses 'with the least possible delay', preferably in the same faculty and monitored by a new and permanent advisory body on education and training set up by the four professional bodies. They suggested that the first year of university education should be spent by all new building, architecture, engineering and surveying students jointly studying building technology, theory of structures, economics, law and management, and history. The Committee regarded the implementation of joint training as an urgent matter for the future of the building industry, stating:

> The specific requirements of architects, builders, quantity surveyors and structural engineers can be met more effectively if students are taught against a unified intellectual background. We are concerned to see that this is done in a joint manner and not separately. Failure by the four institutions to grasp quickly the opportunities for educational development might well lead during the next fifteen or twenty years to the continuation and extension of the divisions in the institutions between those entering the industry after university and those who were trained in a different way.[17]

The *RIBAJ* leader at the time of publication of the report concurred with this view and with the urgency for implementing the changes. The leader went on to blame separate education for 'habits of mind, illusions and methods of work, which lie at the root of many of the difficulties which beset the industry and the professions today'.[18]

Although the Noel Hall proposals were taken up in an experimental way by a number of institutions in the 1960s, they were never supported at an institutional level by any joint body composed of the professional bodies. Nevertheless the attempt to integrate the architect into the building industry, especially in the context of industrialisation, provoked a marked homogeneity in reforms of curriculum content. In 1962, Richard Llewelyn-Davies and John Weeks presented their changes to the curriculum of the Bartlett School after a year in control of the school.[19] The degree course now included the new subjects of anatomy, physiology, psychology, and social geography as well as drawing classes at the Slade School of Fine Art. Reform under Llewelyn-Davies considered the architect part of the building industry exclusively at professional level, with the role of the architectural technician and their separate educational route relegated to 'sub-professional'. This scientific, professional approach to educating the architect effectively distanced architects from low-status manual workers. As far as construction was concerned, Llewelyn-Davies and Weeks wrote:

> We are trying to convert the courses in practical construction from the old rule of thumb approach to a more scientific basis ... we want to treat these subjects very largely as the physics and chemistry of building materials rather than the rules and wrinkles of traditional craftsmanship.[20]

By contrast, at the same time, a few miles away on the Holloway Road, the Northern Polytechnic was proposing a very similar curriculum but with the inclusion of joint first-year training with other building industry trainees.[21] This was argued as essential so that all members of the building team learnt to 'speak a common language'. The proposed course stressed the importance of industrial knowledge for the architect and the ability to design industrial components. The report warned that, otherwise, 'This could well lead to the elimination of the architect as we know him, and a relegation to subsidiary occupations, the most important of which being one of "aesthetic" veneering.'[22]

These reforms to curriculum content can be seen partly as a response to the newly topical interest in 'communication' between members of the building team and an example of new managerial policies entering industry at the time and the research emanating from the Tavistock Institute. They are also part of the social and political upheavals of the 1960s but the fact that none of the suggestions for joint training became institutionalised perhaps demonstrates the deep-seated British class system and the powerful position of the professional bodies in retaining the means to confer professional status. The increased professionalisation of architecture through the reform of training made management, building science, and economics take precedence in reform of educational syllabi, especially so at the Bartlett under Llewelyn-Davies.[23] Although this had the effect of consolidating the position of the architect as a member of the construction industry's professions, it still did not delineate the territory of the profession. Indeed, this was a time when architectural training became increasingly removed from site processes, which paradoxically, coincided with warnings over the distance between design and production.

Operative training in the mid-twentieth century

The defining characteristics of the post-war system of training for building workers were: the distancing of the state from vocational education and training; concentration on apprenticeship in the traditional trades and occupations; reluctance to establish a comprehensive scheme of training outside these occupations; and the constant reaffirmation of the division between education and practical knowledge, between knowledge acquired outside the workplace and within (Cotgrove 1958: 33).

Cole had criticised the ejection of the Board of Education from input to the formulation of the White Paper. This, he anticipated correctly, was an indication of the minimal role educationalists were to play in the organisation of training for young people entering the building industry for the rest of the century. Out of a total of 53 members of the BATC, there were only five members representing educational bodies, including two from the Ministry of Education. The majority of members were from industry, comprising 17 from the NFBTE, 17 from the NFBTO and five representing the different professional institutions. The RIBA, ICE, ISE, RICS and the Institution of Municipal and County Engineers all fielded a member.

No attention was given to the widening of training to take in new and emerging construction occupations or to the problem of having very high numbers of unskilled in the industry. These issues were aired in the industry press but with an air of hopelessness, in that there were no accompanying proposals for changing the status quo. In 1949, an editorial in the *Architect and Building News*, entitled 'Training – for what?' declared:

> The new President of the Reinforced Concrete Association has recently made a plea for more intensive technical training of those responsible for concrete work on building sites – for the raising of the 'concretor' from the status of an 'unskilled' semi-labourer to that of a craftsman . . . he did not suggest anything very definite for the type and place of essential training . . . What practical schemes are afoot towards an all-round improvement? [24]

The answer to the editor's query was, on a national basis, none, and it was only later that the Cement and Concrete Association began running their own short training courses, but these were never accepted as equivalent to training leading to skilled status.

The BATC met regularly and attempted to systematise and standardise apprenticeship training through the setting up of the National Joint Apprentice Scheme for the building industry in 1946 to cover the main crafts, operated throughout England and Wales by Joint Apprentice Committees, consisting of employers, union, education, and government representatives. The registering of apprentices under a standard form of indenture and the issuing of certificates of completion were promoted and administered by the NJCBI on a voluntary basis. This met with only a certain degree of success: in 1951, only 60 per cent of industry apprentices were registered, declining by 1954 to 53 per cent; the rest remained under verbal agreements with their employer. Post-war reconstruction plans, to train 260,000 building trade operatives in six-month schemes in Government Training Centres (GTCs), were initially resisted by the unions, but eventually accepted, on the proviso that they were under the control of the newly established national apprenticeship scheme committees.

Much hope had been pinned on the prospect of nationalisation as a unifying force for the building industry and a catalyst for the reform of training. There had been hopes that a wider technical education for apprentices would be instituted and that training would cover the entire range of occupations in civil engineering and building. But by the early 1950s, with nationalisation no longer on the agenda, the trade unions were aware of the shortcomings of the voluntary agreements of the National Scheme of Apprenticeship and the problems arising from not having an organisation within industry dedicated to the regulation and administration of apprenticeships. In 1954, only

40 per cent of boys entering the industry were indentured but enforcement was impossible without legislation.[25] Some activists thought that the trade union movement itself should take a more active role in education:

> We are, I believe, taking part, in the building and civil engineering industries, in a social revolution ... I would like to see the Trade Union Movement, so far as craft apprentices and manual workers are concerned, giving them more facilities for letting them know the possibilities of town planning, what good craftsmanship really means to our building, to our culture, to our architecture and so on.[26]

The rise of labour-only subcontracting at the end of the decade contributed to the decline in training as it was impossible to include apprentices on gangs where speed was of the essence and bonus systems operated.

The lack of a statutory basis made the whole system difficult to implement and BATC was finally disbanded in 1956, claiming to have established a 'firm foundation' for apprenticeship training in the industry but acknowledging that problems 'remain many and varied.'[27] The 1956 White Paper on Technical Education did outline plans for a massive injection of funds into further education and for major reorganisation. However, until the Construction Industry Training Board (CITB) was established in 1964, training was again left in the hands of the industry where 'custom and practice' masked the ingrained prejudices of both trade unionists and employers (Clarke and Wall, 2010).

Though this was one of the most intense periods of technological and occupational change, it was not until 1963 that the NJCBI commissioned the BRS to run a three-year investigation into building operatives' work on traditional and non-traditional sites throughout England, Scotland and Wales. A total of 5,343 operatives were interviewed, including foremen, electricians, gangers, clerks of works and others, and the results were published in 1966 in two volumes, *Building Operatives Work, Volumes 1 and 2* (BRS, 1966b). The report presented in great detail a trade-by-trade review of the actual work that was being carried out on building sites, both traditional and systems-build. One of the main findings, that the work apprentices experienced on site was entirely due to chance and not the result of any planned scheme of site training, was a clear indictment of the unregulated and unstructured training available in the industry. Another was that, although 75 per cent of carpenters and joiners, bricklayers and plumbers had attended college as part of their apprenticeship, few had gained any qualification. Only 43 per cent of carpenters, 39 per cent of bricklayers and 50 per cent of plumbers held a craft certificate. Finally, nearly half the workforce was found to be composed of non-apprenticeship trades or occupations and thus not in receipt of any formal training whatsoever. The report not only revealed the extent of the training crisis in the industry but also the range of occupations found on sites, numbering over 40, despite industry statistics still only recognising seven traditional building trades: carpenters and joiners, bricklayers, slaters and tilers, plasterers, painters, plumbers and glaziers, and masons (not including electricians and heating and ventilating engineers).

The CITB was set up under the Industrial Training Act of 1964 which represented 'the first attempt to formulate a modern industrial manpower policy' (Perry, 1976: xix). The Act sought to give trade unions a fuller role in training policy through the establishment of tripartite statutory Industrial Training Boards (ITBs), numbering 27 by 1969. This imposed an obligation on employers to train through the institution of a levy-grant system, whereby all firms except those below a certain size paid a levy to the CITB which was then used to subsidise firms providing training, the majority of training taking place in small and medium-sized building firms. In 1957, 75 per cent of all trainees were with smaller firms and this proportion was still high at 70 per cent in 1965, though declining to 63 per cent in 1970. Although the ITBs had very wide remits in the field of industrial training, they had a number of important drawbacks. In the first place, as G.D.H. Cole had predicted, they were not closely connected with the education system. Employers were wary about the degree of coercion that they represented and the trade unions were not consistently in favour either. Indeed the craft unions were generally unenthusiastic as they continued to favour entry through traditional apprenticeship since that route acted as an important control on labour market entry and on the maintenance of pay differentials. The general unions, however, seeing opportunities for their members where none had existed before, were more enthusiastic.

Despite the findings of the BRS research, the CITB did not immediately attempt to overhaul training radically even though it acknowledged that this was demanded, by 'the accelerating pace of technical progress' (CITB, 1966: 7). Instead it developed onsite training schedules for apprentices, gave employers guidelines on standards and their responsibilities towards trainees, and reiterated the habitual mantra of the industry minimising the importance of educational input:

> Traditionally, training for the construction trades has always been on the job, the master or competent operator training the newcomer, whether adult or apprentice. The introduction of widespread day-release studies and training for apprentices during the last 20 years has been a major achievement of which the industry is justly proud ... This is, however, a supplement to training on the job, not a replacement of it. It is widely held that no replacement is possible and against this background the programme of training development has to be planned.
> (CITB, 1967:13)

The CITB did, nevertheless, initiate a long-term programme to review the occupational structure and training needs and in 1969 introduced the New

Pattern of Operative Training (NPOT), a plan of training devised to encompass all areas of building activity. This programme finally attended to the content of training rather than only the length, with the CITB producing all the required training material for use in colleges and other recognised centres and on site (CITB 1969). Despite this high-quality, monitored training, trainees employed by firms paying a training levy to the CITB remained a minority among all construction trainees. The majority of training in the construction industry remained *ad hoc* with either no agreement or only verbal agreements between employer and trainee. There was, however, one area where training reform was pursued: that of supervisory training.

Conclusion

By the time the CITB Board was inaugurated in 1964, there were no architect representatives on the management board. The link between architectural design and its implementation by skilled workers seemed to disappear not only in the establishment rhetoric of industrialised building but also through changes to architectural training. Building workers, and their contribution to the building process, began to slip into invisibility compared to the years following the war when an article on apprenticeship in *The Builder* published in 1949 could address architects as being directly affected by the poor state of industry training.

> To architects, perhaps, the situation is singularly urgent, for hard-won ability at the drawing board cannot adequately be translated into the bricks and mortar on the site without that skill which only the hands of the trained craftsman possess.[28]

For building workers, the separation of architectural education from the rest of the building industry only served to increase the deep divide between those perceived as inhabiting the lower echelons of the industry and architects, described by Richard Coppock in 1950 as inhabiting 'the wonder place in the skies'.[29]

4 Post-war change
Management and organisation

> One outstanding impression that we have gained in our inquiries is that the growing technical sophistication of the construction industry has not been matched by the development of a similar sophistication in the handling of personnel relationships.
>
> *Phelps-Brown Report*, 1968[1]

At the end of the war, the Labour government, voted into power with an unexpectedly high majority, inherited a country on the verge of bankruptcy and with severe problems in its balance of payments. The programme of social welfare reforms on which their manifesto was based and the enormous task of physical reconstruction of war damaged towns placed the new government in economic crisis. The announcement, in 1947, by the US Secretary of State George Marshall that financial aid was available to assist the economic recovery of Europe was greeted with enthusiasm. Britain's 'special relationship' with the United States became closer after a grant of $1,263 million was approved, in 1948, under the Marshall Plan.[2]

The organisation of reconstruction was an enormous task and some in the Labour Party argued for the implementation of centralised labour planning through a differential wages policy. Evan Durbin, Parliamentary Secretary at the Ministry of Works, made a case for a differential wage structure, which would have the effect of attracting workers to undermanned industries, like construction, and argued against retaining the practice of free collective bargaining, which did not endear him to the trade unions.[3] Instead the government opted to promote increased productivity through attempting to arrive at a consensus between the employers and the unions, which meant leaving the machinery for free collective bargaining intact and also refusing to rescind payment by results. Stafford Cripps suggested closer contact with the USA when he initiated the idea of the Anglo-American Council on Productivity (AACP) at a meeting with the Director of the European Co-operation Administration (ECA), the co-ordinating body for the administration of Marshall Aid.[4] For the United States, there were pressing economic reasons for getting Britain and the rest of Europe functioning again as industrial economies but there was also the political objective of defeating Communism and promoting the American way of life as an alternative: two objectives which were actively pursued through the activities of the AACP.

The AACP was a non-governmental body, consisting of trade union and employer representation, set up to promote the productivity drive through a series of team visits by different industries to investigate American industrial methods. This was welcomed by the employers as a way of separating themselves from government intervention in industry, and by the government as a means of escaping the suspicions of the American political administration towards their socialist policies.[5] The reception by the trade unions was not initially favourable. Indeed, the AACP had been set up without consultation with them, and union members were averse to the idea of importing American management practices based on Taylorist ideas of scientific management.[6] Nevertheless, the TUC members co-opted onto the AACP Council soon began to work readily with the employers to create the terms of reference under which the AACP would function. The AACP existed between 1948 and 1952 when it became the British Productivity Council after the cessation of Marshall Aid. During this time it hosted 49 study visits and 17 visits on specialist subjects. All of these were published as reports, with extracts highlighted in pamphlets, on television, and widely circulated in all the industries that had sent teams to America. Dissemination was generously funded via the Marshall Plan.

The actual effect that the AACP had on British productivity is open to question. Tomlinson (*Business History* article, 1991) considers it negligible, with the employers using the scheme as a convenient shield behind which they hid the fact that they had no intention of changing management practice in a seller's market where they were enjoying high profits.[7] Carew (1987), on the other hand, credits the wholehearted support and active promotion of productivity by the trade unions as the means by which scientific management became generally accepted in British industry. These interpretations are only partly relevant to the construction industry and an examination of the report of the building team provides evidence that the employers, especially worried about the effect of full employment on productivity, endorsed scientific management and particularly emphasised the 'individual output' of the operative over broader changes in management practice.[8]

The British government had exhorted workers to higher levels of productivity, particularly in the undermanned construction industry, as part of the shared war effort. With post-war economic policy promising full employment, worker productivity assumed even greater prominence than during the war. Although in the 1940s management strategies were not applied to any great extent in the construction industry, 'the productivity problem' was frequently discussed. As Tomlinson (1994) has pointed out, it was widely and erroneously believed that labour productivity was synonymous with intensity of labour effort.[9] This was especially true of the construction industry and the Ministry of Works, since the setting up of its statistical department, regularly published data on productivity, incentive payments and worker output.[10] The Ministry calculated that in 1947 productivity was 30 per cent lower than in pre-war years but that it improved slightly in 1948 to 26 per cent.[11] The Girdwood Report into the cost of house building also concluded that productivity in the late 1940s had fallen by 31 per cent.[12] It was against this background that, during July and August 1949, the Building Industry Productivity Team made a visit of six weeks to the northern states of America on a mission charged with:

> examining the organisation, constructional techniques and industrial outlook of the American building industry and of drawing conclusions from a comparison of American and British constructional practice likely to increase productivity in the building industry of Great Britain.[13]

The members of the team were heavily weighted towards management and the professions in both numbers and expertise. Although ostensibly representing a joint union–management venture, the employers, arguing that as they and not the unions were contributing towards the cost, frequently selected the employees for the team who were chosen to represent the 'average worker'.[14] Out of the six taking part in the building team, there was only one, John McKechnie a former bricklayer and Clerk of Works who was then a member of the Executive Council of the AUBTW, who had any experience of negotiating at policy level within the industry. By contrast, the six employer representatives included the senior vice-president of the NFBTE, who was also the team leader, the junior vice-president of the NFBTE and the senior vice-president of the Scottish NFBTE, together with leaders of three other employer groups. The four professionals comprised the President of the RIBA, Michael Waterhouse, and Robert Matthew, Architect to the LCC, and two senior members of RICS while the Team Secretary was also Secretary to the NFBTE.

This was hardly a 'balanced' team and given this composition, it is not surprising that the report focuses on the employers' concerns, with both operatives and architects subject to implied criticism in comparison to their American counterparts. In the case of architects, the business-like approach of their American counterparts was commended. It was observed that there was better printing of specifications and working drawings, superior drawing office equipment and filing systems, better referencing and storage of drawings and that complete accounts of office and professional costs were kept. These observations were consistent with the fact that in the United States, unlike the UK, most architects were employed in very large practices (55 per cent of all work was carried out by the 500 largest offices). Emphasis was placed on the differences in the design and contractual process that allowed speedier completion. The team concluded that: 'the early supply of complete and detailed information is a fundamental factor in securing speed of construction'.[15] But the report vindicated the British architectural profession from any real criticism by the inclusion of an attack on the Labour government's controls and bureaucracy in comparison with the lack of restrictions in America.[16]

The building licensing system in Britain, still in operation after the war, was seen as an obstacle to complete pre-planning of work before execution as few clients would risk the expense of the preparation of full working drawings before licence application. Even when the licence was granted, only two months were allowed to elapse before work had to begin, making it impossible, on large schemes at least, for working drawings and bills of quantities to be complete before going on site.

There were, however, few excuses made for the alleged low productivity of the British construction worker in comparison with his American brother. The section of the report on labour productivity opens with a quotation from Alistair Cooke on the competitive spirit of American industrial life:

> You have only to lean out of any mid-town window to notice the furious concentration and energy of construction workers while they're on the job. At five o'clock they will quit like an exploding light bulb, but up to that moment they haul and hammer and drill and bulldoze with fearful zest. If they work this way, they will keep their job; if they don't, they won't. That is the simple, brutal rule of life in America in prosperous times.[17]

This sets the tone for the following section and it appears from the text that there was little input from the operatives in the delegation. Despite the lack of accurate statistics, the absence of any official studies, and a recent statement by the US Commissioner for Labor Statistics noting that there was 'little prospect that we will in the foreseeable future be able to provide a comprehensive measure of production in the contracting industry', the visiting team made their own comparison of productivity between Britain and America.[18] This was based on 'questioning employers and labour representatives to ascertain their impressions of changes in American building productivity since the beginning of the war'. In addition to this, they undertook their own survey and made an estimate of average man-hours for the completion of a number of tasks observed during the site visits they made over the period of the team's six-week tour. These data, though questionable in their rigour, were then compared with the known average British output for similar tasks, with the conclusion that output per man-hour for similar site activities was 50 per cent higher in America than Britain. The explanation offered was that Americans worked at a greater speed than their British counterparts. However, the report did not emphasise the correlation between higher wages and higher standards of living (American construction wages were four times higher than British and higher than any other industry in the USA). Instead, the individual attitude of the building worker was pinpointed as the key to differences in labour productivity. This was coupled with the 'fear motive arising from unemployment' and the existence of a large pool of unemployed waiting to replace the ineffective worker. The report also ignored the high accident rate in American construction due to fewer safety regulations coupled with the intensity of work and the far higher number of man-days lost through industrial disputes.

The final recommendations call upon all members of the British industry to collaborate to enhance productivity. There then followed a list of points detailing changes in architectural practice, site organisation and the lifting of regulations in line with American practice. However, two whole sections of the recommendations were aimed at building operatives in a style that makes it clear why the unions reacted so strongly against the publication of the report. It reflected the adversarial tenor of industrial relations for decades to come:

> Finally, it is necessary to emphasise the vital part, which the individual operative and his union officers must play in the struggle for increasingly higher productivity and progressive reduction in costs. When architect, contractor, subcontractor, materials manufacturer and supplier, local authority and Government department have all made their essential contribution to maximum site efficiency, the ultimate responsibility for production must still rest on the individual operative on the job. The fact that output in the building industry is so much higher in America than in Britain is not due only to the better organisation, which has been developed and the natural advantages that are enjoyed. It depends, too, to a great extent, upon the keenness and initiative of the individual workman, who is proud to be a member of the building industry and anxious by his efforts to maintain the status in society and the standard of living he derives from it. He takes an interest in the job as a whole, and not merely in his own particular operation, and co-operates wholeheartedly with his employer and with the other workmen. Changes in site organisation, where these are necessary to raise output, are readily accepted, and he willingly assists in the development of new methods and techniques.[19]

It is not surprising that the building unions in Britain rejected not just the unfavourable comparison of the 'ideal' American worker with his British counterpart, but the entire report, and had to be persuaded by the TUC General Secretary, Vincent Tewson, to limit their public criticisms.[20] The report was the subject of a long article in the *Operative Builder* which concluded, 'we do not commit ourselves to the unquestioning acceptance of the implications of either the information or the conclusions ... Let America, if it so wishes, pursue its febrile activities ... For our part we choose otherwise.'[21] British building unions considered that the American high standard of living and consumerism were not worth the cost of its four million unemployed and the absence of a welfare system to act as a safety net for the unpredictability of employment. Despite the objections of the trade unionists, 40,000 copies of the report were distributed throughout the industry and four members of the team appeared on a television programme designed to show how American experience could assist housing problems in the UK.[22]

In the same year that the AACP Report was published, the government released its own investigation into the British building industry, *The Working Party Report on Building*, 1950, which made many of the same recommendations as the AACP Report.[23] The government, in fact, found the American report 'of great value in counteracting the damaging effects' of its own investigation.[24] The three professions (architects, engineers and quantity surveyors), were largely cleared of any fault in slowing productivity but architects were marked out as having the most influence on the efficiency of the industry.[25] The report did recommend a common two-year university course for construction professionals before they specialised so that future builders widen their technical curriculum and so that architects increase their technical knowledge and practical experience and bring them closer to industry.

The fall in productivity was explained as due to a combination of factors: scarcity of materials and hold-ups due to licensing, a decrease in levels of skill caused by depletion of men during the war years and a new intake of inexperienced workers which applied to both employers and employees; and a carefully worded 'it is said' critique of workers using the basic, low, rates set under the wartime PBR scheme as the norm. However the 'Attitude of the Operative' was given an entire paragraph which reported 'some evidence of a decline in morale' due to food shortages, inadequate accommodation, the nature of repair work rather than new-build and the reduced fear of unemployment. But in terms of overall industry productivity perhaps the most telling finding was that the decrease in productivity was linked to the current type of work, which was largely repairs to bomb damaged buildings carried out by small firms. Large house building schemes had not been put in place because of economic constraints and there were, up until 1947, large numbers of half-built houses abandoned because of labour and material shortages.[26] Both reports emphasised the importance of management training for the industry to improve efficiency but efforts to set up an industry management college did not materialise.[27] When the reports were presented to Parliament, Richard Stokes Labour MP and Minister of Works, underplayed the productivity statistics, deeming them unreliable and instead focused on congratulating the trade union leaders and emphasising team effort and good management.

> I am determined to do whatever I possibly can to improve production and to get costs down, but, as both Reports say, success depends on better management and on better collaboration all round. Make no

mistake about it, the men are not to blame for low output if there is not good management nor proper collaboration there to help them. The men on the job can, of course, influence results, but they can only influence results in so far as management, organisation and the rest of it serve them.[28]

Stokes, in fact, blamed materials shortages and the inefficiency of private sector firms for the decrease in output, and later in the same debate a number of positive comments were made on the higher output and cheaper houses being built by Direct Labour Organisations of the LCC and other local authority bodies.

Both the Working Party Report and the AACP Report's recommendations on building worker attitudes were considerably different to those of a government wartime visit made only five years earlier. In 1944, Lord Portal, Minister of Works, appointed a committee composed of Alfred Bossom, Sir George Burt (Building Research Board), Sir James West (Ministry of Works) and Frank Wolstencroft (ASW and past president of the TUC Congress) to report on methods of building in the USA.[29] The committee commented favourably on the organisational practices they observed, the introduction of standardisation of building components based on the 4-inch module and the use of factory-produced housing assemblies and they made a large number of recommendations arising from their observations. Most of their recommendations were repeated, in greater detail, by the AACP Report, but with one exception: the speed at which operatives worked. The Portal Report stated that the higher output per man-hour found in the USA was not due to a greater speed of work: 'We saw no indication that the American operative works harder or faster than the British.'[30] Higher output per man was interpreted as due to the co-operation between designers, contractors and operatives, all 'striving' to increase efficiency through better organisation of the work. The team concluded: 'We believe interest and understanding are the most powerful of incentives to work, and we examined American methods for securing both of these on the part of the workers.'[31] The two later reports not only exemplify the rapid erosion of consensus between professionals, employers, and representatives of labour at the end of the war, but an explicit interest in American management practices as a solution to industrial problems. At the beginning of the 1950s building workers were increasingly being represented by employers as obstacles to progress and productivity. There were already by the late 1940s a range of technical innovations which might have increased productivity as seen in trade and architectural press (Figure 4.1) but uptake and investment by private sector firms were minimal, and the industry remained one that preferred to use cheap, and often casual, labour rather than expensive plant. Hod carriers remained a common sight on British building sites until the end of the twentieth century.

The pace of industrialisation was also the subject of intense debate in Europe but with notable differences with regard to labour, compared

68 *Post-war change: management and organisation*

Types of mechanical equipment which speed up bricklaying. 190. *The Chaseside brick hoist.* 191. *The Brickon scaffold jack.* 192. *The Warry platform hoist with automatic safety gates. On the platform is a brick carrier with detachable container, a great improvement on the traditional hod.*

Figure 4.1 Modern plant for brickwork
Source: *New Ways of Building* (1948).

to the tenor of debate in Britain and the USA. The first CIB Congress, held in Rotterdam in 1959, fielded a wide range of papers on different aspects of industrialisation. W. Treibel, describing the position in Germany (GDR) described operative output, or the pace of building, in 'Hours/m^2 of Habitable Surface', while the French contribution was cautious of piece work incentives and instead reported that 'excellent results' had been gained from allowing the workers to keep 'the whole of the gain accruing directly from increased productivity'.[32] Nowhere was the individual worker described, and productivity gains from industrialisation were reported as due to better organisation and increased mechanisation: individual output remained a preoccupation of British researchers explained by a historic and peculiarly Anglo-Saxon understanding of skill as an individual attribute.[33]

Productivity and management

Throughout the 1940s, 1950s and into the 1960s the BRS produced a number of studies on productivity, particularly house building, which attempted to quantify labour input in man-hours and identify the factors that led to the lowest labour input.[34] The techniques of scientific management,[35] in the use of work-study methods, were applied to measure productivity on experimental house types.[36] This research programme compared different housing systems by analysing the man-hours and machine-hours expended by operatives recorded by a 'resident site observer'.

The purpose of much of this research was to find methods of construction that required fewer skilled workers. The first study was published in 1948 with the main objective 'to economise on manpower'.[37] When the second report was published a year later, it was explicit in pointing out the impossibility of defining the ratio between skilled and unskilled in the workforce because there was no commonly understood definition for these two groups of workers.

> The skills required in non-traditional construction may not be regarded as skills in the traditional sense. Nevertheless, skills of the right kind are essential if the best use is to be made of the new methods.
>
> Owing to these limitations in definition, it is impossible to assess accurately enough for valid comparison the requirements of skilled as against unskilled in the various non-traditional types and the traditional type of house.[38]

By 1950, the BRS had given up on attempting to classify the skills needed for innovative construction methods and produced a series of reports that examined new methods of working in terms of man-hours and costs without differentiating between levels of skill.[39]

What these reports eventually revealed was the wide variability between firms and sites and that the labour requirements, *in toto*, were independent of building technique and that even with specialist house building firms, the labour input for identical houses could vary between 900 and 1,300 man-hours.[40]

Donald Bishop summarised the findings of these studies while he was Head of the Production Division at the BRS thus: 'high productivity is a consequence of good design and management in all its aspects rather than the constructional method selected'.[41] The BRS also investigated the effect of wage incentives on building productivity but again, after reviewing the available evidence, Bishop found it inconclusive.[42] He considered that although 'increased productivity may result from direct monetary incentives it is probable that more could be gained from systematic training, by good organisation, and by successful human relations'.[43] Nowhere in Bishop's review of productivity is the individual worker described.

The National Building Studies Reports all emphasised the importance of efficient organisation and administration and the key role of the foreman for high productivity. The employers' organisation, the NFBTE, initiated the setting-up of the National Advisory Committee for the Training of General Foremen in 1950, which produced a number of reports. The Committee included representatives from the Institute of Builders, the Ministry of Education, the Association of Principals of Technical Institutes and the Foremen's Association. The ethos behind the training of foremen, from the employer's perspective at least, had changed with industrialisation. Now that building was perceived as an assembly process and less concerned with issues of craftsmanship, the foreman was 'mainly concerned with the fundamental principles of industrial management and supervision leading through good team-work to satisfactory productivity'.[44] However, there was no integration between the foreman's courses and craft training and this exemplified the lack of integration between the large number of qualifications available in supervisory and management subjects.

From its inception, the CITB had given equal weight to the issues of management and supervisory training in the industry as it had to operative training. The 1967 annual report stated that, 'No aspect of training development has as much significance for the future of the construction industry as that of the supervision of operations and the management of firms.'[45] Statistics on the numbers of white-collar and supervisory occupations in the industry (classified as administrative, professional, technical and clerical) started being collected in 1963 and there was a steady increase in these occupations in the late 1960s relative to a decline in the size of the operative workforce.

Productivity was also a topic discussed by architects, and at a 1963 RIBA Conference on the theme 'The Architect and Productivity', Donald Gibson spoke in favour of negotiated contracts rather than open competition, so that builders could be involved at early stages, and also took the opportunity to promote dimensional co-ordination.

> One of the essential aids to greater efficiency all round is the establishment and use of dimensional co-ordination. You will know that the government is now giving the necessary lead. If architecture offices could support this work by establishing their own vocabulary of recurring details, coded for future use, there could be a great saving in manpower.[46]

Standardising architectural details would, it was hoped by Gibson, free up more time for architects to work on pre-planning, liaising with clients and working on critical path diagrams of the building process. He also spoke on building labour, describing the new techniques in use as demanding new and exacting skills, and looked forward to the day when everyone involved could reap the rewards of being part of an industry that equated with other successful sectors like aircraft and engineering. But only five years later in 1968, the results of a Manchester University study into accuracy and productivity in industrialised building found that communication and organisation were the key points, and also the weakest, in the production process. It concluded:

> We believe that it is significant that a study which was originally intended to determine the interrelationship of dimensional standardisation and tolerances in the field of site erection processes should have its major emphasis on communication systems and organisational problems.[47]

Wages and skill

Modernisers decried the craft basis of the construction industry, blaming the traditional trades as resistant to new working practices, and entry via the apprenticeship system as obstacles to industrialisation. However, there was a steady decline in the number of skilled workers in the industry between 1946 and 1970. Throughout the 1950s and into the early 1960s, the proportion of skilled workers in construction remained at just over 50 per cent, but by 1966 it had declined to 48 per cent and by 1970 to 46 per cent (see Tables 4.1 and 4.2). Some of this may be due to numbers of skilled workers disappearing from industry statistics into unofficial self-employment, or 'the lump' as it was known, but this change also coincided with a decrease in the amount of construction training. There was no doubt this era of industrialisation, and the debates on skill in relation to new techniques, was accompanied by real changes in the structure of the construction workforce.

Increasing mechanisation and the introduction of new materials earlier in the century had given rise to many new occupations; among them concrete

Table 4.1 Proportion of unskilled in the construction workforce, 1946–54 (private sector firms only)

Numbers employed (1000s)	1946	1947	1948	1949	1950	1951	1952	1953	1954
Skilled craftsmen	461.5	529.8	521.5	549.7	550.7	532.7	512.5	514.7	517.1
Unskilled workers	365.2	408.6	428.6	416.8	423.0	430.2	444.7	454.6	450.8
Total	826.7	938.4	950.1	966.5	973.7	962.9	957.2	969.3	967.9
Unskilled as percentage of total	44	43.5	45.1	43.1	43.4	44.6	46.4	46.8	46.5

Source: Ministry of Works Statistics.

Note: 'Skilled craftsmen' includes: carpenters and joiners, bricklayers, plumbers, glaziers, masons, electricians. 'Unskilled' includes categories: 'other building craftsmen' and 'other occupations'.

Table 4.2 Proportion of unskilled in the construction workforce, 1958–70 (private sector firms only)

Numbers employed (1000s)	1958	1960	1962	1964	1966	1968	1970
Skilled craftsmen	577.8	587.8	595.7	611.5	522.1	488.1	403.4
Semi/unskilled workers	477.2	517.3	530.5	542.3	569.3	544.5	461.6
Total	1055.0	1105.1	1126.2	1153.8	1091.4	1032.6	865.0
Semi/unskilled as percentage of total	45.2	46.8	47.1	47.0	52.1	52.7	53.4

Source: MPBW Statistics.

Note: 'Skilled craftsmen' includes electricians, plumbers, bricklayers, carpenters and joiners, masons, plasterers, and painters and decorators. 'Semi/unskilled' denotes all occupations without a recognised apprenticeship, including formworkers, steelbenders, crane drivers, concrete workers, 'Building and Civil Engineering Crafts' and labourers.

workers, steel benders and plant operators, but there had been no parallel change in training to incorporate these new occupations. There had also been a steady lessening of the difference between the wage rates of skilled and unskilled men; the labourer's rate, (about 64 per cent of the craftsman's in 1886) was fixed at 75 per cent in 1922 and raised to 80 per cent in 1945.[48] This has been interpreted as a deliberate strategy in response to technological change by the craft unions (McCormick, 1964). A high labourer's rate meant that there was less likelihood of employers using unskilled and semi-skilled labour instead of craft labour.

With the end of the wartime uniformity agreement, wages for building craftsmen were agreed by the NJCBI, based on the rate for a skilled worker, which covered carpenters and joiners, bricklayers, painters, plasterers, plumbers, and roof slaters and tilers. This payment included allowances for tools, travel time and various other special payments, for example, the holiday scheme introduced in 1942. Members of these trades working on civil engineering sites were paid NJCBI agreed rates, which also included a lower unskilled rate for non-apprenticed labourers working as assistants to craftsmen. On the other hand, civil engineering wage rates agreed by the Civil Engineering Board (later called the Civil Engineering Conciliation Board) were based on the labourer (termed 'general operative') with no tool allowance since there were no apprenticeships and thus no necessary 'tools of the trade'. Supplementing the base rate for the labourer were a range of plus rates paid for various activities, for example, driving a dumper or other piece of mechanised equipment, but payable only for the duration of the time the plant was being used.

By the late 1960s an unskilled labourer's wage was still 80 per cent of a skilled wage but in reality it was possible for labourers to earn more than skilled workers due to the plus rates paid under the civil engineering agreement. This had surfaced on the Festival of Britain site where, although 100 per cent unionised on the building side, the non-unionised labourers working under the civil engineering agreement were earning more than the craftsmen and also encroaching into jobs deemed by the craft unions as their territory, resulting in two strikes and a large number of lesser disputes.[49] This site was unusual as in fact, throughout the 1950s, most demarcation disputes between unions that were members of the NFBTO were settled locally with very few taken to national level for arbitration.[50]

Without a unified training and qualification system linked to wage scales, there was no incentive for many workers to undergo a lengthy, and often patchy – in terms of the quality of training received – apprenticeship when the construction industry offered similar wages for unskilled as for skilled workers. These two completely separate wage systems were a major obstacle to unity in the construction trades (Wood, 1979:177) and a topic of frequent debate within the NFBTO.

In 1957, Richard Coppock, General Secretary of the NFBTO, called for an investigation into the fundamental problems underlying wage differences.[51]

Whereas representatives of other unions, however, in particular, the AUBTW, welcomed the new techniques cutting across traditional trade boundaries as an opportunity for amalgamation. Coppock was unbending in his defence of the craft unions against the encroachment of civil engineering workers. For example, building in concrete was frequently the subject of disputes between the ASW and concrete workers under the civil engineering agreement, both claiming the right to shuttering work at their different rates of pay. By the mid-1960s with the introduction of large-scale industrialised building, the distinction between civil engineering occupations and the building trades became even more opaque. Although the NFBTO Executive had decided, in the late 1950s, that the work involved in the erection of prefabricated buildings was that of craftsmen, there was no clear definition of what comprised labourers' work in building. Many trade unionists feared that wages on the heavy prefabricated systems were based on labourers' rather than craft rates, especially as craftsmen were in the minority on these sites.[52] Again there were calls for a common construction agreement to cover the whole industry as, in the words of one, delegate:

> whether it as a man who lays bricks on top of one another, a man who saws timber or a man who paints, he does it just as efficiently in the Civil Engineering Industry as in the Building Industry, and why should they be employed under two different sets of conditions?[53]

A common construction agreement would have necessitated an integrated training system, for the two conditions were intimately linked. In terms of the acquisition of skill, the craft unions, formally, still argued for traditional apprenticeship as a means of providing the range of experience necessary for the autonomous worker, and the complicated wage system found in the post-war years was a reflection of the unclear status and definition of the skilled or semi-skilled worker and a lack of recognised training for workers in these two categories.

Conclusion

By the late 1960s there was a vast increase in the size of contracts using various forms of industrialised building but no accompanying reform of the training system, and the composition of the construction workforce changed between 1940 and 1970 with a decline in the proportion of skilled workers. The organisation of industry in post-war Britain remained predicated on strict hierarchies with the ethos of command and control the essence of management, especially in construction.[54] Whereas by the 1940s the principles of management were well established in America, and Europe, together with the premise that they were skills that could be taught, in Britain, the idea of 'born leadership' still prevailed (Gourvish and Tiratsoo, 1998). The gulf between manual workers and salaried staff within firms was vast; for example, in the 1960s, Tarmac maintained an executive aeroplane to fly senior managers to meetings around the country. Management hierarchies within firms were also apparent in the different dining rooms at Head Office for different levels of staff, culminating in the top-floor Directors' Board Room and dining room.[55] The idea of worker consultation was anathema in British industry up to the end of the 1960s and particularly so in construction with its history of strong employer resistance to any interference in their affairs, whether from the state or from workers. The hierarchy of the construction site mirrored the hierarchy of the British social class system. The five basic social classes recognised by the Office of Population Censuses and Surveys were described, from 1921 to 1971, as a classification of occupations according to their reputed 'standing within the community'. Throughout the post-war period and up to the 1970s, there was little disjunction between the social class position of architects I (professional), and building workers III (skilled), IV (partly skilled) and V (unskilled).[56] The opinions expressed in the 1950 AACP Report on operatives' 'attitudes' to work were still current two decades later while productivity studies attempted to measure individual output, and it was not until the early 1970s that the more progressive human relations approach to management began to appear in construction.[57]

Part II
Architectural abstraction
The role of the Modular Society in promoting industrialised methods

5 The Modular Society

> [I]t was British architects above all others who, in the post-war years, threw themselves into the 'industrialization of building', and by their enthusiasm carried the process further than architects had ever done before.
>
> Reyner Banham, 1968[1]

Introduction

The Modular Society was an indispensable element in the part that architects played in the industrialisation of the post-war building industry. Sir Alfred Bossom, MP, and former architect, and Mark Hartland Thomas founded the Society, in 1953, to promote industrialisation 'by co-ordinating dimensions of materials, components and fittings on a modular basis'.[2] Membership of the Society spanned the building industry to include manufacturers, architects and contractors and provided a valuable, and rare, forum for discussion and communication. Describing its membership as comprising, 'all those concerned with building from clients to craftsmen', there was, however, a noticeable and significant absence of any representatives of the operative workforce.[3] The debates and arguments surrounding many of the key determinants of industrialisation, the size of a module, joints and jointing methods, tolerances and many others, were aired at Modular Society meetings, recorded in the transactions of the Society and published in the Society's journal, *The Modular Quarterly*[4] (Figure 5.1).

The records of the Society's transactions start in 1953 on mimeographed sheets that were privately circulated. By mid-1955, the Society had established itself sufficiently to publish number 13 of its transactions as the first issue of a glossy journal, *The Modular Quarterly (MQ)*. From 1963, the proceedings of the newly formed International Module Group were also included in *MQ*. In 1970, with declining membership and shortage of funds from advertising revenue, the Modular Society accepted an offer to incorporate *MQ* into *Official Architecture and Planning (OAP)* with Mark Hartland Thomas joining the editorial board as Technical Consultant. This lasted until April 1972 when *OAP* changed direction and became *Built Environment*, the first issue under the new name consisting of an in-depth

80 *The Modular Society*

Figure 5.1
Cover of The Modular Society flyer, c. mid-1960s

Source: Bruce Martin's papers.

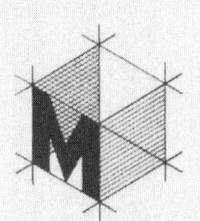

THE MODULAR SOCIETY

Join The Modular Society and keep your modular know-how up to date

THE MODULAR SOCIETY LIMITED
22 Buckingham Street, London, W.C.2.
TRAfalgar 4567

look at the Department of the Environment with a leading article by Dame Evelyn Sharp. Mark Hartland Thomas left the editorial board of *OAP* (at the same time that Bruce Martin joined it) and the transactions of the Modular Society were again published independently as a four-page leaflet under the title *ModSoc News*.[5]

The *Transactions* reappear, in 1977, in their final incarnation as a photocopied four-page leaflet called *Module*, which existed until 1980. The penultimate edition in 1979 gives an account of the Modular Society's twenty-fifth Annual General Meeting with 'a mere 25 members in attendance'. On this occasion, Lord Beeching retired as President and in his final address pointed to the dangers of the Society having achieved its objectives early, so that it no longer had a useful role. Even though this was hotly disputed by the few remaining members, the next edition of *Module*, which appeared in June 1980, was the last. Ironically, this final edition carried an account of the speech made at the Annual Luncheon by John Bennett, then Head of Construction at Reading University, in which he referred to recent research into the claimed benefits of modular co-ordination (by this time known as dimensional co-ordination). This piece of research, initially instigated by the Modular Society and funded by the Science Research Council, reported that as construction was affected by so many different factors 'simply re-structuring the dimensions on drawings' had very little effect on what happens on site.[6] This was perhaps a simplification of the aspirations of those involved in the early years of its activities, but a critical reminder of their failure to fully examine site processes.

The Modular Society campaigned for the universal dimensional co-ordination of construction components. This, it was argued, would reduce the wastage involved in traditional building, as modular structures would be built to order and fitted together on site. In turn, this would transform the construction process, from muddy chaos to clean and rational orderliness. The idea was a logical progression from pre-war prefabrication methods but instead of using a large range of different systems, the Modular Society promoted rationalisation through a unified system of components, which would be chosen by the architect from a catalogue and fitted, ideally without any adjustment, on site. As such, modular co-ordination was promoted in Britain and internationally as a system:

> [which] places at the disposal of the individual creator accurate and universally understood definitions which make it possible both to reduce the diversity of the product used in building – and to manufacture them on an industrial scale – and to reconcentrate in the hands of the designer the power to define *in toto* the object to be achieved.[7]

Clearly aimed at architects, this rhetoric was an attempt to defuse the resistance of many towards the industrialisation of building and possible restriction of their design freedom by emphasising their increased power of control over design. Modular co-ordination, later termed dimensional co-ordination, was a key aspect of industrialisation and a concept, as promoted by the Modular Society, that highlighted the position of the architect as central to both design and organisation. It is not surprising then, that architects dominated the membership of the Society outnumbering all other categories of membership.[8]

The following account attempts to illuminate the ideas underpinning dimensional co-ordination through an appraisal of the work and careers of two leading members of the Modular Society: Mark Hartland Thomas and Bruce Martin. They respectively espoused two of the common architectural responses towards industrialisation: from Mark Hartland Thomas, the unassailable right of the architect to direct proceedings, and from Bruce Martin, an equally unassailable belief in the power of rationalisation to shape the process of architecture.

Mark Hartland Thomas and the Modular Society

Mark Hartland Thomas made the inaugural Alfred Bossom Lecture at the Royal Society of Arts in December 1952 the occasion where he presented the 3 feet 4 inches (40 inch) planning module, as a basis for introducing modular co-ordination to British building practice.[9] Based on 'the shoulder-width of a human, plus the tolerance for movement plus constructional thickness on either side to centres of support', it was 'fundamentally right because it is securely founded upon human scale, the most enduring factor in architecture'.[10] Hartland Thomas continued:

> But human scale has its importance in manufacture as well, and more still in assembly, for it controls what a man can lift and handle. The economy of machine-production, with ever larger and more powerful machines, demands the production of larger pieces: limitation to human scale ensures that they are economical to handle on the job.[11]

This aspect of human scale revealed the way that he conceived modular components in relation to the workers who were to build with them. The statement implied an aversion to, or ignorance of, the heavy pre-cast systems that were soon to make their appearance in Britain. It assumed small-scale assemblies with a low level of mechanisation – more like a craft-based production process than large-scale highly mechanised industrialised sites.

The human scale argument was, however, incidental to Hartland Thomas' main assertion: that modular co-ordination would result in cheaper buildings through economies gained from the mass-production of standardised building components. These building components were to be based on a 4-inch module.

The 40 inches planning module (or grid) integrated well with the American 4-inch module ratified by the American Standards Association in

1945. Up until the mid-1950s, Britain was the world's largest exporter of prefabricated houses, with exports to the value of £5.5 million in 1951 and £6.9 million in 1952, mainly to Canada and Australia.[12] A module that fitted into the imperial system of measurement was therefore going to be of far more immediate use to British firms relying on the export market than a metric-based module.

In the closing remarks of the Alfred Bossom Lecture, Mark Hartland Thomas suggested that a private society be set up to 'promote research, experiment, development and discussion concerning the use of the module in the design and construction of buildings and in the manufacture of building materials'. Sir Alfred Bossom seconded this and early in the next year, on 6 January 1953, a statement to the press on the aims of the new society was released:

The Statement to the Press[13]

The object of the new Society will be to contribute towards lowering the cost of building by co-ordinating dimensions of materials, components and fittings on a modular basis. At present we are not getting full advantage of flow-production because standard and customary sizes of different components do not fit together. Recent experience in the schools programme and in house production for export has shown that the use in design of a human-scale planning module of around 3 feet 4 inches is economical of time and money, and it influences the production of components to come into phase with the module.

The Modular Society will afford a common meeting ground for architects and other technicians, contractors and manufacturers to pool their experience in modular planning in order to solve some of the difficult problems still outstanding and to extend the vocabulary of modular components to designers over the whole range of building construction (brickwork included) instead, as at present being chiefly limited to a small number of proprietary systems.

In this way the Modular Society will supplement by direct industrial experience any government research work that may be initiated under DSIR and the study of modular planning already underway at the British Standards Institute. The society is not likely itself to engage in research except to a limited extent in special cases.

Quite apart from the subject, the range of membership, bringing together the technical, contracting and manufacturing sides is likely to be an important new influence in the building industry.

Alfred C. Bossom LL.D FRIBA MP
M. Hartland Thomas OBE MA FRIBA

The first meeting, by invitation only, was held at 5, Carlton Gardens, SW1 on 19 January with 19 attendees.[14] The Society was made immediately financially viable by a donation of £250 from Alfred Bossom (Figure 5.2) and a Provisional Committee inaugurated to create an infrastructure. Further discussion decided that although the bulk of finance should come from industry donations, it was important that professional and industrial members should have equal status. It was also decided that the Society should emphasise that it was equally concerned with 'wet and heavy' as well as 'light and dry' construction.

Between January and October 1953 the Modular Society held four private meetings of the Provisional Committee. The design of the headed notepaper was decided on, 'Mr Bossom preferred a dignified notepaper heading in copper-plate script', and a logo by Austin Frazer of the Design Research Unit was chosen as the symbol to represent the Society.[15] At the final provisional meeting on 6 October, Mark Hartland Thomas reported back on the motion he had successfully presented at the International Union of Architects Congress in Lisbon which recommended the adoption of the 4-inch module.

With the Modular Society up and running, the Provisional Committee was dissolved and in November 1953 the first elected Executive Committee appointed Mark Hartland Thomas to the post of Secretary, at a salary of £500 per annum, a position he held until his death in 1973. Over the intervening 20 years he became identified as the public face and driving force behind the Modular Society. The years preceding his appointment as Secretary to the Modular Society had provided Hartland Thomas with a number of experiences that gave him the credibility and position to found the Modular Society. These included: as a theme convenor for the Festival of Britain, as the leader of the Ministry of Works Team to assess the post-war German building industry, and as Chair of the RIBA Architectural Science Board Committee on dimensional co-ordination.

Mark Hartland Thomas' architectural career spanned the middle of the twentieth century, from the prefabrication and rationalisation movements of the 1920s and 1930s to the massive technical changes of the 1960s. On his death in 1973, the *AJ* obituary described him as 'a large man, bowler hatted and booted, he was not everyone's image of the architect, except for his quiet sincerity and high sense of purpose and duty – a salt of the earth character'.[16] Monica Pidgeon, editor of *Architectural Design* throughout the 1950s, remembers him as 'a charming giant of a guy who ran the Modular Society'.[17] He described himself, in an autobiographical sketch in the *AJ* in 1959, as liking Beethoven, Brahms, dancing and *Panorama* and the enjoyment of driving his grey, 1934 Daimler.[18] Born in 1905, he gained first-class honours in classics at Cambridge in 1927, before studying for his architectural qualifications at the Royal West of England Academy, Bristol. In 1932, he became a partner with his father P. Hartland Thomas, the Diocesan Surveyor for Bristol, and designed two houses in 1936 and 1937 respectively. One of these houses appeared in the *Architectural Review* of December 1936 as part

Figure 5.2 Lord Bossom examining a Modular Society questionnaire while on a site visit
Source: *Modular Quarterly*, Autumn 1960.

86   *The Modular Society*

of F.R.S. Yorke's review of 15 other Modern Movement architects.[19] After this he produced very little architecture but did collaborate with Felix Samuely in 1945 on the design of a space frame for roof construction patented as 'Unitectum'.

He became a member of the MARS (Modern Architectural Research) group in 1936 and acted as Secretary in 1944–48. In 1940, he took the post of Deputy Chief Architect with United Dairies Ltd., where, because he fulfilled the criteria for being in a reserved occupation, he remained throughout the war years. During the war he was also a member of the Design and Industries Association (DIA) Council from 1942–48 and Chairman of the London Region between 1944 and 1946. After the war, between 1946 and 1951, he held the post of Chief Industrial Officer at the Council of Industrial Design (COID). In 1952, he became 'an independent consultant' and devoted the rest of his life to working for the full acceptance of modular co-ordination in Britain.

Hartland Thomas had been actively involved in research since 1945, when, as Chairman of the RIBA Architectural Science Board, he was asked to lead a team from the Ministry of Works to assess the German building industry at the end of the war. The team, which included Richard Llewelyn-Davies, toured the west of Germany between October 1945 and January 1946, and was one of many organised under the umbrella title of British Intelligence Objectives Sub-Committees (BIOS) sent on intelligence gathering operations to Germany. The report resulting from this visit contained a substantial section on standardisation and the use of dimensional co-ordination in wartime German building, obtained through the translation of relevant documents and a series of interviews with Ernst Neufert who had been appointed Commissioner for Building Standardisation in 1942 by Albert Speer.

All wartime building in Germany had been regulated by dimensional standardisation set down in the German Industrial Standard (DIN) No. 4171. According to the BIOS Report, this system had been devised by Neufert and derived from 'the gradual development of building types' and analysis of existing building components.[20] Neufert decided on 2.5 metres as the base unit for grid planning in the frame construction of industrial buildings and 1.25 metres for living accommodation. The system was compulsory under the Nazi administration, and it continued in use even after the Nazi ordinances had been abolished, suggesting that it had become, to a certain extent, standard practice. Whether it had provided economic benefits was a question on which the team was going to interrogate Albert Speer, detained at the time of the visit in Kransberg Castle, which housed most of the senior scientific and technical personnel of the Third Reich before they were removed to Nuremberg.[21] It was here that Speer, and others, were interrogated by both American and British officers and civilian experts. If this interview ever occurred, there is unfortunately, no account of it in the report.

The British assessors dismissed the German grid sizes and questioned the validity of the theoretical basis for Neufert's system. They concluded that, although it was supposedly derived from the existing sizes of standard door and windows, the actual reason for the choice of the grid dimension was arithmetical, based on subdividing 1000 into halves, quarters and eighths to achieve the relative sizes 500, 250 and 125. This gave a dimensional series that fitted conveniently into the metric system. Neufert had expanded his ideas into a book, *Bauordnungslehre* (1943) that the British report harshly reviewed as a 'product of the German professor who has to write a thick book for reasons of prestige'.[22] They concluded that Neufert's thesis was undermined by his admission in conversation that measuring ancient buildings could prove anything, and it was therefore rejected as having 'the flavour of *ex post facto* reasoning'. However, after describing the shortcomings of Neufert and his book, the team concluded that 'the importance of his dimensional system as a subject for study by British experts is undiminished'.[23] A copy of *Bauordnungslehre* was removed for careful scrutiny by a 'leading British expert on dimensional standardisation' who would then 'make recommendations for adoption in Britain and America'.[24]

The 'leading British expert' was the RIBA and the subject of dimensional co-ordination was handed to Hartland Thomas as Chairman of the RIBA Architectural Science Board (ASB). He promptly set up ASB Study Group 3 to focus solely on the subject of dimensional co-ordination. Although he was later to become a leading protagonist of the Modular Society's campaign to introduce the 4-inch module as a basis for standardisation, Hartland Thomas initially disagreed with the BIOS team's position on German dimensional standardisation. He perceived the rejection of the German system as short-sighted and expressed his own opinions in his book *Building Is Your Business*.[25] Short, at 119 pages, but well illustrated, this book is typical of a number written during the late 1940s in the informative but somewhat didactic style of the professional expert introducing modern themes in architecture to the layman.[26] Here, in complete opposition to the official view, Hartland Thomas actually praises Neufert's planning grid and rejects the 4-inch module:

> A system that had great success in Germany during the war, and is still in favour there, laid down a plan-grid based upon 1.25 metres as the unit, with no control over heights. All buildings had to be designed on this grid, which acted like a magnet to influence the sizes of ready-made components to come into phase with the grid, making what allowances were necessary for tolerances and for thicknesses of other members.
>
> This German system has very interesting possibilities and it has a better chance of success than the American or British researches which are aiming at the establishment of a small unit (3 or 4 inches) applied in all three dimensions, heights as well. Such a small unit would not realise economies sufficient to offset the disturbance caused to production by adopting it; and the standardisation of heights, as well as horizontal dimensions, would put a severe encumbrance upon architectural design.[27]

He appears to have been generally impressed by the German construction industry and includes praise for the 'subtle and complicated' German system of PBR whereby gangs of skilled workers graded their own capabilities and shared the bonus proportionately.[28]

The official views of the BIOS Team were wholeheartedly supported by Walter Gropius who had been invited back to Germany at the end of the war by General Clay to advise on whether to continue with Neufert's system for post-war reconstruction. His opinions on the origin of Neufert's 1.25 metre module were later quoted at length. He considered it should be abandoned because:

> the module on which the whole system was built up in Germany was not 'biologically' derived but was based on purely mathematical roots, taking about four feet as the basic module. I consider this dimension unsuitable as it is too large for an entrance, window or staircase in minimum residences – and double that dimension is too long for a bathroom or for a bed.[29]

Gropius believed that the basic dimension used should be related to human proportion. He took the width of an 'average' man as a basis, giving 3 feet 6 inches as the ideal, but in his own practice this had been modified to 3 feet 4 inches to fit in with the American Standard 4-inch module. This was the grid dimension adopted by Gropius and Wachsmann in 1943,[30] in their ill-fated attempt to manufacture prefabricated timber panel houses in the USA with the General Panel Corporation.[31]

The complete dismissal of Neufert's work owes something to the very high levels of anti-German feeling at the end of the war. Gropius' comments are particularly harsh, but perhaps understandable as Neufert was a former pupil who continued to work under the Nazi regime. Neufert was one of the first students at the Bauhaus in 1919, leaving to work in Gropius' private practice between 1924 and 1926.[32] He had been developing his system of standardisation for many years and in his first book on the subject, *Bauentwurfslehre* (1936), clearly states that his system is based on subdivisions of the lengths of the human head, face and feet after the work of Dürer. He also acknowledged the nineteenth-century work of Zeising on the dimensional relationships of human proportions and the later research of Moessel. This information is given under a half-page illustration based on Dürer, including a diagram of the golden section. Entitled *Der Mensch das Mass Aller Dinge* (Man is the Measure of All Things), this page has survived virtually unchanged, in every subsequent edition of Neufert's *Architect's Data*.[33] Neufert's illustration bears a striking resemblance to pages in Le Corbusier's *Modulor* published ten years later in 1946.[34] Neufert, true to his Bauhaus education under Gropius and Hilberseimer, carried on his investigations into the rationalisation of the building process throughout the war years. His imaginary *Hausbaumaschine* of 1943 is an extraordinary exploration of the idea of industrialising the process of house building using a moving factory, consisting of scaffolding on tracks, to construct terraces of houses. The BIOS Report described it as a 'monstrous building machine . . . of doubtful practical value', and Jean-Louis Cohen has described it as an example of a 'perverted modernism . . . [used] in the service of oppression' and one that typified architectural expression in the Third Reich.[35]

The British team was especially interested in how the entire German building industry had adapted to the imposition of modularity: an experimental idea with many disciples in both Britain and the USA where it was implemented on a voluntary basis.

The American Standards Association had recommended a 4-inch module in 1941 as a basis for co-ordinating all sizes in the building industry.[36] However, on mainland Europe, it was rapidly moving into legislation as in the French 1942 Standard, NF P 01-001, recommended that dimensions of both buildings and components should be multiples of 10 centimetres (10 cm).[37]

Meanwhile, the British were considering a number of proposals. In 1945 *Post-war Building Study Number 2, Standard Construction for Schools* recommended a module of 8 feet 3 inches. However, in the same year, the Ministry of Works published *A Survey of Prefabrication* which recommended a dual approach of using both 4 inches and 40 inches.[38] This was an attempt to reconcile the conflict between the small and large module by using 40 inches as a planning grid with components sized in increments of 4 inches. This architect-authored report was a detailed and beautifully presented review of both pre- and post-war prefabricated methods of housing construction. The authors were clearly in favour of modular co-ordination as a means of increasing production on an industrial basis:

> The prefabrication movement has been groping towards a goal of modular co-ordination without being able to attain it. Our analysis has led us to the view, amplified in the text that a basic dimensional standard is a prerequisite for satisfactory prefabrication on a nation-wide scale . . .
>
> The proposal we have put forward in some detail . . . is for a compound Modular System of either, 10 cm–1 m or 4 inches–3 feet 4 inches. This we believe to be sufficiently flexible to embrace most of the materials likely to be used and most types of building.[39]

But early in 1946 the Ministry of Works Standards Committee published a report, which contradicted these recommendations. In *Further Uses of Standards in Building*, the 4-inch module was rejected as inappropriate as it was related to the size of American building components, and a 3-inch module, related to the size of a British brick, was proposed as the basis from which component sizes should be derived.[40] This report was also negative about the use of modular planning grids, considering them 'a rigid factor dominating design, planning and construction' and, according to Hartland Thomas, 'made many of us very angry'.[41] The MoW Standards Committee was large, consisting of 43 members, chaired by Sydney Tatchell, an architect known for his traditional approach, with 14 of the other members also architects with a traditional outlook.[42] The government-published, official history of the introduction of industrialised methods of building supports the stance taken by the Standards Committee.

> The Committee believed that agreed dimensions for components, as advocated throughout its work, would provide the advantages of standardisation, while leaving the designer free to assemble them with greatest economy. To adopt the same modular basis for all purposes would not only dislocate production during the change-over period, but would entail the immediate rejection of a large proportion of existing moulds, patterns and machinery, thus placing an unnecessary burden on industry without corresponding advantages. This would be especially true of the post-war period. [43]

The Standards Committee report took a cautious and pragmatic view of what was possible for the British construction at a time of extremely high demand. The possibility of creating a *tabula rasa* on which a new and co-ordinated system of modular components was to be manufactured was not an option considered economically viable.

Although both the construction press and popular press frequently published articles on the housing problem and its possible solutions, mass production of building components was only one option. Throughout the 1940s and 1950s the rationalisation of traditional brick-built housing was also being investigated in government-funded research at the BRS. In the opinion of many Modular Society members, it was precisely this government-funded programme of research at the BRS that caused an impasse on an official recognition of modular size that lasted for 20 years.

The RIBA Architectural Science Board's Study Group 3 on Dimensional Co-ordination, which included two of the authors of *A Survey of Prefabrication* and Richard Llewelyn-Davies as chair, first convened on 10 February 1947.[44] The group worked in close consultation with the BSI Committee on Standardisation in Building chaired by S. Rowland Pierce. The RIBA group published their findings two years later in June 1949, however, nothing appeared in the *RIBA Journal* until April 1951 in order to coincide with the publication of BS 1708; 1951, when neither group gave a definitive recommendation for a module size.[45]

Before the study group finished its researches, however, Le Corbusier had produced his own contribution to the debates on standardisation and modular building in his *Modulor* theory, first aired in Britain at the sixth Congrès Internationaux d'Architecture Moderne (CIAM) conference at Bridgewater in 1947.[46] Mark Hartland Thomas provided an account of the conference for *Architectural Design*, a journal for which he was a regular contributor from the late 1940s and throughout the 1950s writing on new American architecture. In a private letter to Stamo Papadaki, a fellow member of CIAM preparing to publish the *Modulor* in America, Mark Hartland Thomas was critical of Le Corbusier's work.[47] He voiced doubts on the usefulness of a system based on what he considered an arbitrary, as opposed to a scientifically rigorous, choice of 6 feet for an average human height and thought that its infinite number of variables was unsuitable for the industrial manufacture of co-ordinated building components.

The RIBA ASB Study Group 3 had collected evidence from a limited range of sources representing only certain sectors of construction, indicating that modular co-ordination was being conceived as associated with only certain types of construction technology: 'light and dry' was favoured over concrete systems. Submissions from the schools building programme dominated, with contributions from William Tatton-Brown and Bruce Martin in Hertfordshire, and Stirrat Johnson-Marshall and David Medd representing the Ministry of Education.[48] Examples from the housing sector included Uni-Seco, the wartime producer of prefabricated houses, the Arcon steel bungalow and the Riley-Newsum House, a prefabricated bungalow based on a 3 feet 4 inches grid and exported mainly to Canada: all three of these systems were designed by architects. The Wates house (the only concrete-based contractor system reviewed) was included as an example of a closed system with a few standard plans that would benefit from modular co-ordination.

The final report suggested that the reason progress towards dimensional co-ordination in the building industry was slow was because there was no financial incentive – the loss incurred from innovation falling ultimately on the community or owner of the new building. It was therefore dependent on a public or private body with a large-scale and long-term building programme to promote dimensional co-ordination and not the responsibility of the industry. The Ministry of Education's school building programme was cited as exemplary, where a planning module was used as the basis for the systematic manufacture of the required building components. Bruce Martin, anticipating future debates, annotated his copy of the report with the comment 'systematic industrial production of building components that are inter-related depends on the design of connections and not necessarily the use of a dimensional module in planning'.[49]

The ASB Study Group concluded:

> Any system using only a single dimension with its multiples and sub-multiples will give rise to architectural proportions based on only the double square etc. This would obviously be an intolerable stultification of design. Freedom of aesthetic expression is an essential requirement which must be met. It seems that it might be met by the development of a series of preferred sizes related to the basic dimension but not simple multiples or sub-multiples of it, such sizes being related by means

of one of the mathematical series used in classical and renaissance architecture.[50]

Despite the Study Group including three architects, Hartland Thomas, Jessica Albery and D. Dex Harrison, openly in favour of a 40-inch planning module and 4-inch component module as a basis for modular co-ordination, the conclusions do not endorse this approach. Instead the recommendations reflect the interests of the chair, Richard Llewelyn-Davies, and the reference to a series of preferred sizes refers to Le Corbusier's *Modulor*.[51] This was welcomed by many architects as an aesthetic, and thus thoroughly architectural response to the complex problems of standardisation (see Chapter 7). It also reflected the preoccupation with proportion, which had been a constant feature of debate in the architectural press since the 1930s.[52] Ernö Goldfinger praised Le Corbusier's attempt to relate a system of proportions based on the golden section to human dimensions and from there establish two geometrical progressions, in turn related to one another. However, he perceived the main difficulty as due to the sizes of components generated from the geometric series failing to fit with door heights, door widths and room heights. According to Goldfinger, even though the *Modulor* was:

> not yet the complete answer to Modulor planning or standardisation, it is nevertheless an enormous step forward and certainly more applicable than the vague conceptions of 3 feet 4 inches grid bandied about loosely by so many incompetent Modulor planners.[53]

The 1951 BSI Report also dismissed the development of modular co-ordination based on a small-sized module of 4 inches (10 cm) as 'misconceived' and suggested that further study should focus on development of a planning module of 40 inches. Over the next ten years the BRS researched 'preferred sizes' through the exploration of mathematical series as a basis for co-ordination, while the BSI in a team led by Bruce Martin investigated dimensional co-ordination in relation to building standards. Differences of opinion over the application of these approaches resulted in the acrimonious meetings and journal articles detailed later.

As well as an accumulation of knowledge and experience on modular co-ordination through research, Hartland Thomas had also acquired relevant practical experience while working with Hugh Casson on the Festival of Britain. While still employed at the COID, he became a member of the Festival co-ordinating team, theme convenor for the downstream section and originator of the Festival Pattern Group where all the participating firms produced designs based on crystallography. The task that had most relevance to his later work for the Modular Society was that of compiling the stock list of all the manufactured products exhibited and used in the Festival of Britain. It was to become the basis of the Design Council's design index and was the inspiration for the *Modular Catalogue*. In 1952, he headed another investigative team, this time under the aegis of the Anglo-American Council on Productivity.[54] Although the visit was limited to the engineering industry, Hartland Thomas managed to continue his personal research interests by meeting with Konrad Wachsmann in Chicago to discuss standardised space frames.[55]

Mark Hartland Thomas summarised these investigations for his inaugural Alfred Bossom lecture, including the range of current architectural positions on modular co-ordination, and in summing up looked at the implications for design.

> Hitherto architectural design has begun with a plan of a certain size and subdivided it into its parts, creating a self-contained composition. Modular architecture proceeds in the opposite manner: it is additive instead of subdividing: one unit is added to another and it matters very little where you stop . . . But after all, towns develop in this additive fashion and the public may well find that modular architecture produces a quieter more harmonious civic design . . .
>
> For architects it is more difficult to swallow: it seems to deny their birthright as designers . . . In architecture we strive to give Savile Row service at a reach-me-down cost. No wonder it is said that only bad architects can be successful. Modular design might allow the good ones to be successful and the bad ones to be good. Also: it could arrest the blind-alley movement towards standardised complete buildings which is anathema to architects.[56]

At the end of the lecture the comments from the floor gave an indication of the range of views on modular co-ordination. Herbert Manzoni, chair of the Building Division of the BSI, thought that the main task ahead was one of propaganda; first, to convince architects of the benefits of modular co-ordination and, second, the government, for without a substantial building programme, manufacturers were not going to take the risk of changing to modular sizes.[57]

Richard Llewelyn-Davies spoke against the concept of grid planning, having found it not essential in the application of modular co-ordination to traditional building, and instead praised the work of Le Corbusier. He considered that work should proceed in the field of a related series of sizes, which were in turn related to some single dimension as yet undecided. Ernest Hinchcliffe, managing director of the firm Hills, makers of prefabricated steel frames and windows for school building, presented the viewpoint of the manufacturer:

> Four years ago my firm went into prefabricated school construction based on eight foot three modular planning, which was the recommen-

dation of the Ministry of Education. My firm has been associated with about five million square feet of construction on this basis. Now this module is being discarded in favour of three foot four inches. We have spent a lot of money, both in development and capital outlay for eight foot three modular planning. We appreciate the position and we are moving on to the three foot four inch grid, but this matter must become firmly established before firms can lay out considerable sums of money on this type of production.[58]

The differing viewpoints aired at the lecture indicated the urgency, for many, of reaching a working solution to the problems of dimensional co-ordination. In January 1953, at the inaugural meeting of the Modular Society, 111 members signed up on the spot.

6 'Additive architecture'
The early years of modular co-ordination

> Design – that mysterious word! – is only disposing elements in proper ways and arranging how work should be done.
>
> William Lethaby[1]

Bruce Martin also had extensive experience of designing and building using methods based on grid planning and modular units.[2] Martin started an engineering degree in Hong Kong but on returning to Britain, in 1935, began studying architecture at the Architectural Association. His first month as a student coincided with the arrival of E.A.A. Rowse as principal, and Martin enthusiastically entered into the new educational framework instigated at the AA where students worked in units to encourage teamwork, utilising a scientific approach to design problems using sociological analysis and town planning.

With the outbreak of war, he spent five years working for Short Brothers, the aeroplane manufacturers. Here he continued his engineering training, attending Northampton Technical College alongside the apprentices where he learnt how to operate lathes and the other machine tools found in the engineering workshop. After the war, Martin joined the Hertfordshire school building team under Stirrat Johnson-Marshall. He worked for the first few years using the 8 feet 3 inches grid system and produced one of the earliest examples, working together with Mary Crowley and David Medd, on Cheshunt Junior School, finished in 1947. His frustration with the unwieldy large grid was eased with his appointment, together with Anthony Williams, in 1948, to develop a new approach, based on 3 feet 4 inches. This smaller grid and the materials later used at Summerswood were first trialled on the Clarendon Secondary Modern School at Oxhey, built in 1951. Summerswood School, finished in 1953, was entirely Bruce Martin's design and also the final outcome of the development group's work.

Although Martin had acquired practical experience and knowledge of the building process, his approach to design was abstract rather than pragmatic. While at the AA, he and David Medd had both been inspired by a talk given by Alvar Aalto in which he referred to the 'domination of unnecessary variation'. After the war, when Medd and Martin began working at Hertfordshire, one of their first achievements was the reduction in range of steel stanchions used for the schools' structural framework. Bruce Martin was also a member of the ASB Study Group on Standards, and it was in this field that he was later to devote the main years of his career. Martin named Lavoisier, the eighteenth-century French chemist, and Geddes as his chief guides in his work on classifying and standardising building components. Patrick Geddes,[3] trained as a biologist with T.H. Huxley, was a polymath whose interdisciplinary approach to city planning became influential with architects working in the mid-twentieth century. His holistic approach to understanding cities, including the processes of repair, renewal and rebirth involved the constant interplay of many factors over place and time. In an attempt to graphically display the complexity of his ideas and theories, Geddes resorted to the use of diagrams that he termed his 'thinking machines'. Lewis Mumford, his most famous disciple as well as his most rigorous critic, criticised Geddes' obsession with his 'thinking machines' as a betrayal of the most original and liberating of aspects of his philosophy – that of 'life's organic interconnectedness' and 'humankind's capacity for insurgence'.[4] Mumford interpreted the diagrams as traps, set ironically by the very master and 'merciless foe of all closed systems' in that they precluded any new developments emerging from their closed system of 36 squares.

Geddes' idiosyncratic graphical method does succeed in displaying some of the complex relationships between different phenomena governing the growth of cities, and it is understandable that they were both intriguing and appealing to architects used to graphical communication as a *modus operandi*. For Bruce Martin, they provided a rational framework from which to start the mammoth task of identifying, classifying and standardising the myriad components found in the building industry, and he refers to Geddes in his 1971 publication, *Standards in Building*.[5] However, he also struggled with Geddes's framework in relation to describing the interaction of the building process with form as the following prose-poem written in the late 1940s reveals.

> Postulate:
> The only permanent form is process
> The only static thing is dynamic
> The Geddes pattern is consequently valid as a static
> demonstration of a dynamic process
> ...
> 'Form in change' can be shown diagrammatically.
> The process is the form (within change). It is the permanent
> phenomena within a moving flux. Process is not change
> itself. It is a state of being within change.
>
> Bruce Martin, 24.3.48[6]

Poetry, in this context, seems an appropriate medium for grappling with the huge difficulties of attempting to unify the complex processes described by Geddes. It could be interpreted that the wider context of the upheavals of post-war life and the determination by Martin and his colleagues to build differently, efficiently and more cheaply also form a backdrop to the poem. However, it is also an attempt to make architectural sense, in the poem's preoccupation with 'form', of the new building techniques that Martin had been grappling with in Hertfordshire. He wanted to revolutionise the entire building process by creating an architecture of parts, to be assembled on site but at the same time was anxious that form should not be eclipsed by process. Poetry was a medium also used by Buckminster Fuller, the design guru deeply admired by Martin, in *Untitled Epic Poem on the History of Industrialisation*, which ran to 226 pages and was also written in the 1940s.[7]

Martin's knowledge of prefabrication and modularity, therefore, was already advanced by the time Mark Hartland Thomas proposed the establishment of the Modular Society in 1952. His comments from the floor at the December lecture at the RSA were not related to the argument over module size but referred to the establishment of a reference system, similar to navigation, so that architects could decide which, if any, modular components should sit on the grid lines and he was already aware that the critical problem ahead was that of joints and jointing.[8]

1953: the Modular Society's first year

Given the conflicting reports that emanated from the Ministry of Works and the Ministry of Education, it is not surprising that the early meetings of the Modular Society were dominated by discussions on module size. The first three public meetings of the Modular Society, held in March, April and May of 1953, and recorded as *Transactions Volume 1*, were debates on the theme of modular co-ordination. The first meeting started with a general discussion opened by a panel of speakers consisting of Mark Hartland Thomas, Stirrat Johnson-Marshall, Chief Architect to the Ministry of Education, the engineer F. J. Samuely, Donald Fraser, Managing Director of Troy and Company Ltd., and W.A. Balmain, Joint Managing Director of Uni-Seco Ltd.[9]

Johnson-Marshall asserted that he and his colleagues at the Ministry of Education had no vested interest in any particular module and were investigating four systems – three based on a 3 feet 4 inches module and one based on a 4 feet module. At the moment the best size appeared to be 3 feet 4 inches which had also been recommended by the RIBA and the BSI. Felix J. Samuely pointed out that standardisation and planning grids were not connected, as had been suggested in the Ministry of Works standards report. The size of standard units differed with the material used and so the best option was to use a small module that could not be subdivided.

W.A. Balmain considered the most compelling reason for modular co-ordination was the reduction of costs. Previously, labour had been cheap and materials dear, now that the reverse was true, efficient use of labour on site or in the factory was essential. The debate had been opened by a contractor, Donald Fraser, who had also referred to labour as being one of the major difficulties in introducing new methods into building. He blamed the 'objection and resistance of tradition-bound craftsmen ... reluctant to accept new principles since their skill was their main asset and they feared unemployment.'[10]

Two further public meetings were held to elicit views from members, with discussion stimulated by questionnaires circulated by Hartland Thomas. These revealed a wide range of views, some of them fiercely held, and the fourth meeting, on 7 May, was the occasion for announcing the results of 39 completed questionnaires and debate the proposition to adopt a 4-inch/40-inch frame of reference as a working hypothesis.[11] Roger Walters spoke against. He queried the whole nature of a 'working hypothesis' in an industrial as opposed to a laboratory setting. 'It was questionable whether the Society could safely adopt a major proposition of the nature proposed, involving as it did the whole building industry and then discard it later.'[12] He suggested that the Society face up to the differences of opinion expressed, and collaborate with both the BRS and the BSI, not risk coming into conflict with them, and urged the meeting to reject the motion as 'hasty and ill-considered'.

Walters' motion failed and the meeting resolved to set up two study groups. Study Group 1 on *Interchangeability*, convened by Hartland Thomas, was to begin the publication of the *Modular Catalogue*, assuming a 4-inch/40-inch frame of reference. Study Group 2 on *Principles*, convened by W.A. Balmain, was to work on the theoretical framework upon which a modular system should be constructed and applied. At the end of the meeting, William Tatton-Brown tactfully suggested that the discussions might be clarified by the examination of existing modularly designed buildings (see Appendix 1). This was taken up and the next meeting was held, in June 1953, at Summerswood School with Dan Lacey, Bruce Martin and Anthony Williams as guides.[13]

Bruce Martin had designed Summerswood School, with Anthony Williams as his assistant, during 1953, and it was the culmination of his research work in Hertfordshire. Based on the 3 feet 4 inches tartan grid, he was aiming to produce a new system, which would be 'aesthetically and proportionately superior to what had gone before, and at the same time fully flexible'.[14] Modular Society members toured the school with Dan Lacey as a guide, after which he introduced them to the thinking behind the design in a short talk. He explained the uniqueness of Summerswood in that it was the first time that the fixed equipment inside the school, that is, the WCs, lavatory basins and cloakroom fittings, had been designed within the grid lines of the plan. All the major structural components related to the planning grid, but by placing the partitions of the smaller rooms on the centre line of the grid, all non-standard detailing had been avoided. This was made possible by the removal of the stanchions from the grid lines and centralising them within each square of the grid at 1 foot 8 inches.

Martin, in his presentation to the visitors, chose to describe the sequence of erection of the building that had been 'assembled rather than constructed'.

> While the services were being fitted, the wall panels were delivered to the site in a furniture van and were unloaded by two men, carefully carried onto the floor slab and stacked. Panels were complete in every respect: glazed, fitted with ironmongery and stove-enamelled. The outside [Holoplast] panels were put up around the building and the interior then seemed like a large department store, with men fixing the ceiling panels and laying the floor screed and floor finish.
>
> ...
>
> The internal panels were put in position on the finished floor towards the end of the job at the same time as furniture, light fittings, and other final components. It was interesting at this stage to sense the atmosphere of the job. I remember two L.C.C. foremen responsible for work on traditional houses visiting the building. The interior was quiet (the sound-absorbent panels ceiling panels and the cork floor being in position). Men were dressed in white overalls, wearing plimsolls and using screwdrivers, ring spanners and pliers. There was no dust, no dirt, no waste material, no broken bricks, no 'building' in the accepted sense. The foremen commented on the quietness and the way each man was getting on with his job.
>
> ...
>
> It has been suggested that I tell you of the difficulties, complications and things that went wrong, but this is difficult to do as the building was put together part by part without hindrance.[15]

This account encapsulates Martin's approach to building. Although when the school was opened, the teachers disliked the 'off-grid' columns which broke up the classroom space, and there were problems with the joints, architecturally, this school is credited with elegant proportions: the deep clear lines of the roof fascia which held the services foreshadowing school architecture of the 1960s and 1970s.[16] Over 40 years later, in conversation and recalling the years he spent at Hertfordshire, Martin commented that the skills required for building the schools had resided in the men working in the factories. He still maintained, with an attitude common among engineers that the skills and accuracy of the workshop were superior to those found on site. The Hertfordshire school building programme had relied on medium-sized local building firms. These were used as general contractors for the schools, employing 'a few labourers' to lay the concrete base with a carpenter called in because he was regarded as more competent than the others. It was the manufacturers' men who erected the steel framework, installed the plaster partitions, and laid the woodblock floor. Martin also considered that it was these very same small building firms, after gradually re-establishing themselves over the post-war years, which managed to contrive an end to the lightweight prefabricated schools by offering to build more cheaply in traditional brick.

'Modular Discord'

At the beginning of what was intended to be a new language of building, there were inevitably misunderstandings and arguments over terminology. Differences of opinion were soon aired in the architectural press and the *Architects' Journal* gleefully headlined two articles, the first by Bruce Martin, followed by a response from Mark Hartland Thomas, 'Modular Discord'.[17] Bruce Martin argued against the current concept of modular co-ordination because he considered it resulted in closed systems and instead suggested a method of co-ordination that gave an open system, to allow the development of 'families of components'. He considered it impracticable to demand all building elements and components to be produced with dimensions related to one module as this failed to recognise the realities of thickness, jointing, structural loading and the relation between these. It also presupposed the existence of an ideal building system that would be discovered once the various practical problems had been solved. For Martin, with his practical experience of school building, closed systems had already been found to be perfectly adequate for certain types of building but he baulked at the idea of extending these systems to all buildings of any type.

He concluded by recommending the use of interchangeable 'families of components' that were dimensionally co-ordinated within each family and interconnected by an *in-situ* or easily cut material. He argued for a 'truly architectural conception of co-ordination i.e. co-ordination by planning, not by forcing manufacturers to alter the dimensions of their products to "co-ordinate" with an, as far as they are concerned, arbitrarily chosen module . . . '.[18] This would place the role of design and organisation of the building process entirely, and properly, in the hands of the architect. This article was prefaced by a 'Glossary of Building Terms' written by Martin in order to clarify the new nomenclature of modular co-ordination.

Mark Hartland Thomas's reply, printed immediately next to Bruce Martin's article is a terse response to the publication of the glossary. He was well aware that this had been compiled by Bruce Martin and his response is an unpleasant piece of writing, criticising the author for defining terms that are still under discussion and not ready for 'general consumption'. [19] Hartland Thomas' piece is written in a condescending manner, the overall effect being that of the older man 'pulling rank' over the younger for stepping out of line and publishing information that Hartland Thomas obviously considers he has the right to veto. In particular, he disagrees with the definition of 'module', defined by Martin as a unit that can be the basis for either a planning grid or a building component. Instead Hartland Thomas insisted that:

> It is more convenient and productive of clearer ideas if 'module' is taken to mean the smallest increment in dimension that is recognised in a system of modular co-ordination. This has both etymological and ancient usage on its side. *Modulus* is in Latin the diminutive form of *modus*; it is a *small* unit of measurement. This accords with its classical use as the half diameter of the column at its base for setting out the proportions of an Order. If we say that the module is 4 inches and any component whose overall dimensions are properly related to a multiple of 4 inches is accepted as modular, then we avoid the confusions caused by arbitrary sub-divisions of a larger unit.[20]

However, Martin's definition echoed that published by the BSI in its first report on modular co-ordination in 1951, where the module is defined, open-endedly as 'a unit to be used in the design of buildings and in the co-ordination of design and construction and of standard building materials and components'.[21]

Very soon after the publication of these two articles the post of Head of Modular Co-ordination Studies was advertised at the BSI. The two authors applied and Bruce Martin was successful, a result he attributes partly to the position he took in the *AJ* article quoted above. This early dispute seems to have been completely forgotten by Bruce Martin and in conversation he revealed no malice towards Hartland Thomas, describing him as a clever scholar who knew his Greek and Latin and could stand up and deliver a good lecture. Their paths crossed frequently in the following years, both professionally, as one of the strategies of the Modular Society was to field as many members as possible to sit on the various BSI Committees, and socially. Bruce Martin remained at the BSI for 30 years while Hartland Thomas committed himself to the Modular Society for the rest of his life, which under his guidance became an early type of consultancy to the construction industry.[22] The dispute over the meaning of terms, however, was pale in comparison to the vitriolic exchanges in response to the position the BRS was to take on modular co-ordination over the next decade.

Theory into practice: the Modular Assemblies

The Modular Society formally adopted 4 inches as a basic module size in May 1954, on the recommendation of the newly formed Study Group 2 Principles. This decision, made very early on, enabled the compilation and publication of a Modular Catalogue of components compatible with 4 inches that were currently available from manufacturers. Although certain members, including Bruce Martin, had favoured a more cautious approach advising more research before a final decision on module size was made, Mark Hartland Thomas took the view that:

> You cannot found a new society and immediately go into recess, whilst another organisation sorts out all answers to problems. Besides, we did not consider dimensional co-ordination to be primarily a subject for research, but rather for development combined with public relations.[23]

He forged ahead with plans for the Modular Assembly, proposed in 1956 and built in 1958. Promoted as the first building to be designed entirely on the basis of the 4-inch module it proved a very successful public relations exercise for the Modular Society.

The first drawings for the proposed Modular Assembly appeared in the *Modular Quarterly* of Autumn 1956 drawn up by Hartland Thomas who was eager to experiment with the idea of designing to strictly applied modular rules.[24] He was disappointed with the 'so-called' modular details received from America which appeared to be 'nothing more than conventional details done on squared paper' and wanted to see what would happen if the grid lines were made to really coincide with the boundaries of the components.[25] Building on the previous work of the Modular Society Technical Committee on joints and tolerances, recent work by Bruce Martin on component size, and acknowledging the findings of the European Productivity Agency Project, which had been published earlier the same year, he designed a 20-feet cube.[26] The structure incorporated as many changes of material as possible while still maintaining 'unity of idea' through strict adherence to three design rules:

1 Every component, even structural framework, must keep its station on the assembly-grid (only fastenings, weatherings and trim may cross grid lines between components).
2 Modular components must have appropriate external dimensions in whole multiples of the module measured to centres of joints.
3 A modular joint must only occur within the clearance between two components.

The third rule, a definition of a modular joint, is followed by the caveat that the actual size of clearance in modular design is, as yet, undecided. Hartland Thomas suggested that, 'As a rule of thumb I would venture the opinion that clearance will always be something less than ½ inch . . . ' He goes on to emphasise the distinction between the modular co-ordination of components – the 'manufacturer's business' – and the design of modular buildings – 'the architect's business'.

The difficult problem of jointing was encountered in the early stages of design. The space between components taken up by the thickness of joints displaced the grid leading to the necessity of making a half-module joint in order to 'keep station' within the modular grid. Hartland Thomas excuses this disjunction, plainly occurring on the south elevation, explaining that 'such a limited admission of the half module does *not* mean that a new module of half the size has been adopted'. This would have compromised the underlying theory of modular co-ordination that all modules are *whole* multiples of the smallest reasonable unit. Naval terminology was used to describe the role of the architect, the designer being 'the commodore of a convoy'.[27]

After publishing the draft plans for the Modular Assembly in the Autumn 1956 edition of the *Modular Quarterly*, Hartland Thomas handed over the design to the Technical Committee for comment and improvement. Amended plans drawn by R.A. Sefton Jenkins, structural engineer, appeared in the Winter edition, together with an invitation to Modular Society members to participate in the erection of the Assembly on the basis of 'provide and fix'.[28] The response was 'electric' with 26 firms initially expressing interest, including Pilkington Brothers, Hills, Holoplast, Gardiner, E.C. Gransdon, Howard Farrow and Crittalls.[29] The full list is a roll call of firms already familiar with the school building programme in Hertfordshire and well known to that particular architectural community. They were named in the next edition of *MQ* together with an advertisement for vacant slots available for specialists needed to complete the Modular Assembly.[30]

This was successful in recruiting the firms necessary to supply the missing parts, and in the Spring of 1958 the Technical Committee printed a congratulatory piece on the success of the 'paper-exercise stage' of the project. Eighty different modular components from 36 different manufacturers had been detailed into the tight space of a 20-feet cube.[31] The next stage was the on-site assembly of the small building (Figure 6.1), although the Technical Committee carefully pointed out that, actually, it was not meant to be judged as a building.

> The Modular Assembly is not a building. An attempt has been made to give it a passable appearance, but interest is invited solely in its demonstration of the possibilities of building with modular components, indicated in this experimental Assembly.[32]

The only experimental aspect to the assembly was dimensional; the techniques of building and jointing were all traditional, as were the materials used. Non-modular materials consisted of paint, mastic, mortar and render, glass and felt and sub-modular fixings, and bolts and screws. The object was to use as many modular components as possible that could be bought 'off the shelf' to demonstrate the possibilities of an open approach to dimensional co-ordination rather than the closed systems, like CLASP, that depended on manufacturers supplying specific components for each respective scheme.

104 *The early years of modular co-ordination*

Figure 6.1 Cover of *The Modular Quarterly*, showing the first Modular Assembly
Source: *MQ*, Autumn 1958.

The professed object of the Modular Assembly experiment was to explore tolerance and buildability problems in modular building but it was also excellent publicity for the Modular Society. Hartland Thomas, the designer and driving force behind the organisation necessary to get the building completed, with the assistance of Alan Diprose, enhanced his working drawings into exhibition drawings. These were published in full in *MQ* Autumn 1958 (see Figure 6.2), including a double-page exploded axonometric: a drawing form usually associated with engineering drawings, assuming a factory-based manufacturing process, and conveying the fitting together of all the various components (Russell, 1981: 383). Despite Thomas' assertions that the Modular Society was not entering into empirical research, his *modus operandi* was that of a scientific inquiry. He devised a tabulated report form to be used by participating firms to record the experience of working on the Assembly in order to collect a set of data that could then be analysed. As soon as the building was completed, a detailed survey was planned to assess how accurately the various components fitted into their modular spaces and a series of post-mortem public meetings were arranged for dates following the dismantling of the Assembly.

At a meeting held at the RIBA on Friday, 9 May 1958, representatives of the participating firms agreed to contribute £36 each towards the cost of building the Modular Assembly on land donated by David Carter of the Carter Group.[33] Work started on site on 1 July, but, as was usual in the late

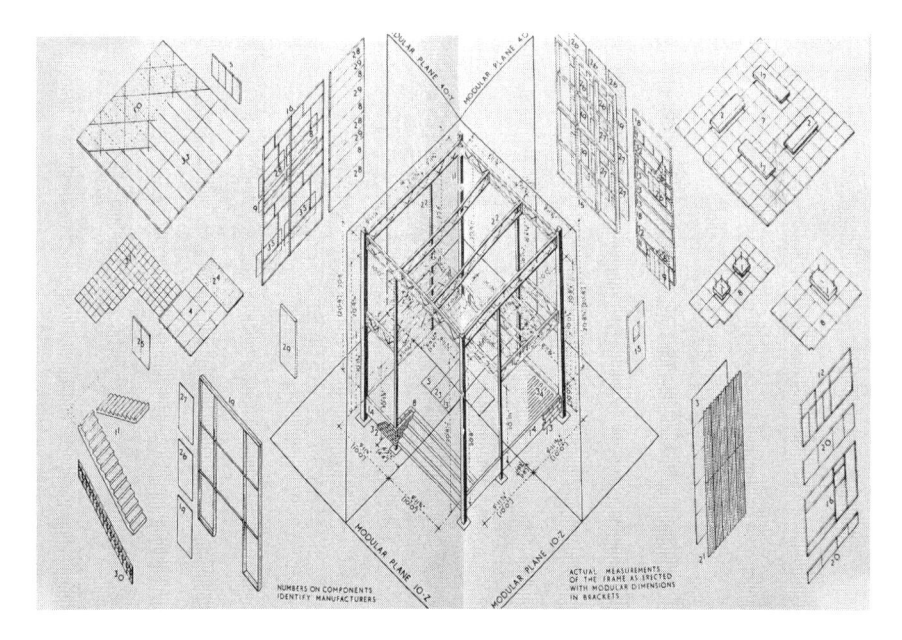

Figure 6.2 Presentation drawings for the first Modular Assembly
Source: *MQ* Autumn 1958.

106   *The early years of modular co-ordination*

1950s, supplies were unreliable and there was a delay of 14 days before the delivery of the structural steelwork. The building officially opened on 1 October 1958, and the Modular Society footed the cost of employing an uniformed commissionaire to man the door. He appears, in a publicity shot, holding open the door of the Modular Assembly and 'keeping station' behind Mark Hartland Thomas standing in the foreground (Figure 6.3). Less

Figure 6.3 Mark Hartland Thomas on the steps of the first Modular Assembly
Source: RIBA Photograph Collection.

than a month later, negotiations were in place with the architects of Hemel Hempstead New Town Corporation with a view to them taking on the Modular Assembly and rebuilding it as 'a pavilion in a public park or similar use'.[34] It was dismantled on 16 December 1958. It succeeded, during its short life, in attracting attention to the efforts of the Modular Society to transform the building industry; Hugh Molson, Minister of Works, visited on 7 November and HRH Prince Philip on the 26 November. Reports appeared in the trade press, the *RIBA Journal*, *The Builder*, the *Architect's Journal* and *Prefabrication* but, despite the royal visit, the scheme did not make *The Times* or surprisingly, considering Hartland Thomas' long connection with the journal, *Architectural Design*.

Post-mortem

The first Public Forum on the Modular Assembly was held on 24 September, before the official opening, at the building centre with Peter Trench acting as a congenial chair and directing the meeting as a question-and-answer session.[35] Mr Pryor, of Howard Farrow Ltd, answered questions on site work, even though he informed the meeting that he had supervised 'from the office'. He did, however, think that introducing engineering tolerances on site was 'going too far, particularly in view of the shortage of skilled craftsmen in the industry after the war'.[36] One of the site general foremen added that in general the main problem was 'ensuring proper tolerances'. Pryor rephrased his criticism of using tolerances too fine for site work: 'It was one thing for an engineer with a theodolite to set out precisely on the grid line, but another for a crane driver to place a 3–4 ton unit precisely on the point indicated.'

For the manufacturers, Peter Gardiner, maker of metal windows, announced that he did not think modular co-ordination would reduce the costs of building unless manufacturers were prepared to produce for stock so that goods were ready for delivery when required. His experience of supplying windows for the school building programme had shown that it was only pre-planning, started several months before work started, that reduced costs. The opinion of the brick manufacturer who had supplied a small number of modular bricks was even more negative:

> Changeover to modular bricks throughout the country would be complicated because the industry was divided between pressed bricks made on fixed dies on machines and the wire-cut principle – on which the modular were made – which unfortunately only produced about 2 per cent of the total output of the industry. The cost of changing over the press machines would be enormous.[37]

However, Hartland Thomas reassured the brick manufacturers that ordinary bricks were easily incorporated into modular brickwork and that modular bricks were not essential for modular design. He pointed out that the Modular Assembly itself held an example of the flexibility of brickwork. It had been noticed that 10 feet above the ground floor, a brick joint was 1½ inches out of true with the modular plane, but over the next 5 feet the bricklayer had lost the extra inch and half to bring the joint back to coincide again with the modular plane. This was taken as an example of the great flexibility inherent in brickwork, even in the most difficult dimension, height.

The roof slabs, although manufactured to the required negative tolerance of minus ⅛th inch when laid in the usual way measured overall 19 feet 11⅞ inches, which resulted in gaps at both ends abutting the column and also against the perimeter beam. This was due to the structural steel framework being out of true. The operative relaid them all close-butting and managed to create a larger gap at one end that was then covered by an extra steel channel. While this solution:

> neatly overcame one of the site difficulties, but it has resulted in all our under-tolerances adding together to leave a gap and at the same time has almost covered up just one of those problems which experiments with a modular assembly should have laid bare.[38]

There were further problems with all the close-butting components: floor tiling, ceiling panels and Vitrolite wall linings supplied by Pilkingtons. Tom Markus (ARIBA) was the Pilkington representative and explained that the glass Vitrolite panels had all been cut ⅛th inch larger than the modular size, probably because the workmen were confused by the notation used to describe the tolerance. However, it was later admitted by Hartland Thomas that part of the problem was that he had detailed a finely jointed material like Vitrolite directly into the coarser fit of the main structural framework. The manufacturer of the ceiling panels, which also did not line up with the modular grid, argued that they had to be close-butted for technical reasons – large joints would lead to pattern staining.

Perhaps the most fundamental of the problems encountered, and the one with the most consequences, was the inaccuracy of the structural framework. Here it was found that not only were the dimensions of the components inaccurate, but the reinforced concrete columns had also been positioned out of their 'stations' on the modular grid, leading to further cumulative problems in fitting adjacent components. There was discussion on the practice of setting out, Mr Tomlin of Howard Farrow suggesting that in future it would be more accurate to commence setting out from the centre line of a building than to start at an edge. Bruce Martin recommended the practice of using packing shims to make the frame assembly coincide with the grid. Mr Pryor again commented on the reality of site work:

> It was one thing for the purist designer to pinpoint exactly the position of the column but another to do it in practice when handling a 20 feet

long column in a high wind with a large crane. In fact about 1½ inches tolerance was allowed on the base plate bolts for adjustments, but it was impossible to pull or push a column weighing a couple of tons to within ⅛th inch.[39]

Although problems with the construction had been conscientiously noted and were reported back to the technical committee, they were overlooked at a special meeting held on 1 December 1958 to assess the experiment. Here only two problem areas were considered: the difficulties of fixing finely fitted components directly within the structural frame and the problem caused by cutting the panels $1/64$ inch larger than the module size. Furthermore, on the general question of accuracy of construction carried out by the building operatives, it was suggested that:

> whilst we should not demand greater accuracy in assembly than is reasonable current practice, the adoption of modular co-ordination was likely to produce an automatic and progressive improvement in accuracy. This had been the experience in mechanical engineering over the last 150 years.[40]

How this 'progressive improvement' was supposed to come about was not discussed and, despite the considerable problems with the building process, Hartland Thomas and the other members of the Technical Committee announced the experiment an unqualified success:

> After discussion it was agreed that it would be opportune to say that there is no need to wait for further refinements in modular method, but that modular buildings should now be designed in full confidence that modular components will be forthcoming and with tolerances settled by their manufacturers according to the method that we have shown.[41]

This assessment was published in the Spring edition of *Modular Quarterly* for 1959, together with guidelines on 'How to Begin' advising architects and builders on techniques for using a 4-inch module grid, which had anyway already been decided on and published in 1958 (Figure 6.4).

The architectural press was more circumspect. The *Architects' Journal*, while generally praising the activities of the Modular Society warned, 'that all the problems of modular co-ordination have not yet been solved and it would be a great disservice to further progress if the Modular Assembly were taken as an indication that they had been'.[42] This article also pointed out that the discrepancies between the different components were very noticeable to the eye and that the partitions stopped short of the ceiling, allowing the ceiling panels to run over and avoid being cut. This was not possible in real buildings where the partition might be load-bearing or have to go through the ceiling to act as a firestop or acoustic barrier.

THESE ARE THE
5 ESSENTIALS
OF MODULAR COORDINATION

```
  4   8  12  16  20
 24  28  32  36  40
 44  48  52  56  60
 64  68  72  76  80
 84  88  92  96 100
104 108 112 116 120
```

BASIC RANGE OF MODULAR SIZES

COMPONENTS (sections, units, compound units)

4 in.

SIZE OF MODULE

MODULE GRID

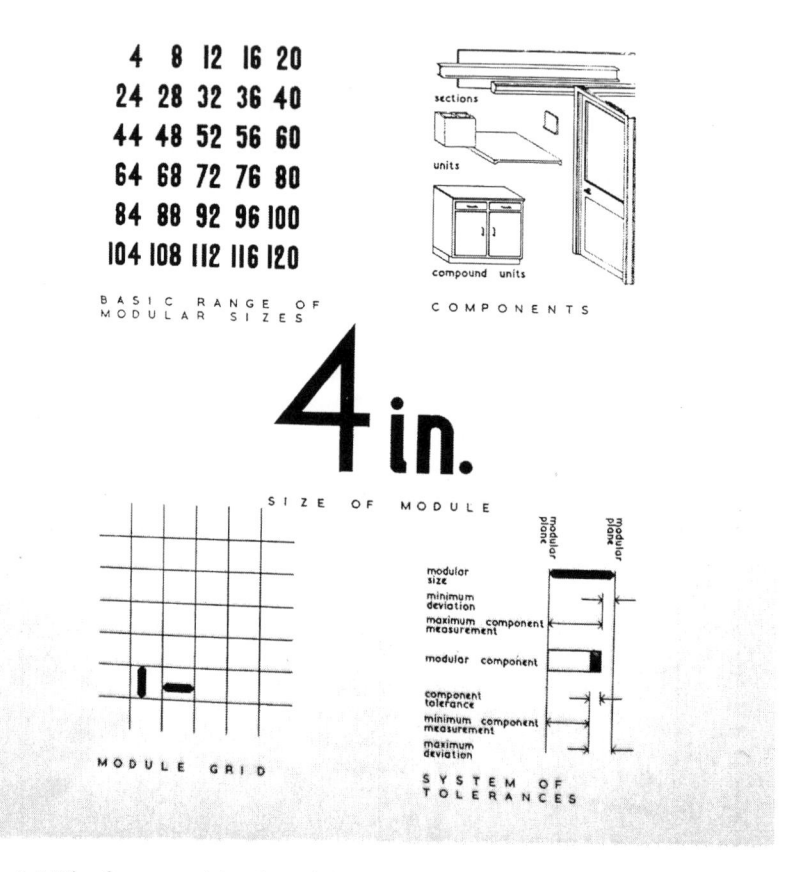

SYSTEM OF TOLERANCES (modular plane, modular size, minimum deviation, maximum component measurement, modular component, component tolerance, minimum component measurement, maximum deviation)

Figure 6.4 The five essentials of modular co-ordination, a summary of five years' work

Source: *MQ*, Summer 1958.

The optimistic summing up of the Technical Committee did not represent the views of all those taking part in the project. Michael Hatchett, a Modular Society member and site foreman, was responsible, for a short time, for the supervising of the construction of the Assembly; he remembers the site erection as being chaotic and poorly organised, with a number of changes of site agent during the construction process.[43] He also recalled that the actual construction tolerances were quite different to the design tolerances and that as the contractors were delivering their services for nothing, in fact they were paying for the privilege of being involved, the project was understood by them purely as a publicity exercise. This is a long way from the type of controlled scientific experiment Mark Hartland Thomas expounded in his correspondence and lectures as the foundation of a new type of architectural practice suitable for the modern world of rational technology.

Despite the statement on the inside cover of *The Modular Quarterly* that the Society's membership included all those concerned with building 'from clients to craftsmen', there were no members representative of industry operatives. Their views were not sought or seen as contributing any necessary information to the new language of building being formulated in Modular Society meetings. However, not all members of the Society had a complete lack of interest in the skills required for building in a modular fashion. Members with extensive practical experience of non-traditional building methods were more concerned with actual site practices. In 1956, in an address to a meeting of 50 Modular Society members, William Tatton Brown drew on his experiences of school building in Hertfordshire to explain the new problems arising from building with prefabricated components:

> It is often supposed that insofar as mechanisation eliminates manpower, it reduces the need for human skill. In fact, of course, the reverse is the case. Prefabrication calls for a greater and not a lesser degree of skill at all levels, and it might not be out of place here to record some paradoxical conclusions which we have arrived at in common with many others who have employed the use of prefabrication. First of all on the job, the craftsman has to exercise a higher degree of care in handling costly components of far higher quality and finish than those normally delivered on to a building site. A far higher degree of precision in fixing is also required as the slightest bodge or botch will be immediately apparent.[44]

He also pointed out the need for manufacturers to produce elements to high tolerances and concluded that, as far as site skills were concerned a new trade of 'Assembler and Jointer' was needed, who would be 'equipped with hand-powered tools, toggles and rivet guns and all sorts of mechanical devices which enable industrial products of high precision to be assembled quickly and accurately.'

Modular Assemblies 2 and 3

By 1964, the subject of industrialised building (IB) was uppermost in contracting, government and architectural circles and publications. The NFBTE

112 *The early years of modular co-ordination*

organised the first IBSAC Exhibition at Crystal Palace in July to promote the sale of proprietary housing systems to local authority clients. The Modular Society decided to participate to demonstrate the use of modular co-ordination in an architect-led alternative to both local authority consortia and contractor closed systems. Still pushing the benefits of open as opposed to closed systems, the GRID Method, developed by Morrison and Partners, in association with the Lancashire County Council's Architect's Department for school building, was chosen as the constructional system for the Modular Society's pavilion (see Figures 6.5 and 6.6).

The GRID structural components of pre-cast concrete were interchangeable with each other and, when designed on a modular grid, allowed modular components from a range of manufacturers to fit into modular spaces. The components themselves, however, were designed with their own proprietary joints and were not compatible with structural components from other systems – strictly speaking, therefore, GRID was a closed structural system just like the others on the market. Alan Diprose was the job architect and the pavilion was erected by A. Monk and Company, who also made a film of the building being constructed. Unlike the first Modular Assembly, there were no post-mortem meetings to analyse any problems with dimensional tolerances as, according to the architect, there were few. Providing minus tolerances were used, that is, that the finished sizes of components

Figure 6.5 The second Modular Assembly at IBSAC 1964
Source: *MQ*, Autumn 1964.

Figure 6.6 The second Modular Assembly exhibition of modular components
Source: *MQ*, Autumn 1964.

were slightly smaller than the respective, nominal, modular size, there were few problems of fit. Alan Diprose, infused with the optimism of the 1960s, regarded the construction industry as being in a transitional state between traditional and industrialised building. He suggested that for the time being, tolerances between components should be relatively large and 'progressively to reduce them as site assembly becomes more mechanised and the element of human error reduced'.[45] The use of a single structural system and textured concrete cladding panels gave the pavilion a visual coherency lacking in the first Assembly. The detail drawings, on modular graph paper, were also confidently presented and there were no aspects of the building that gave any sign of it being an experimental structure. The *Architects' Journal* praised it as 'an excellent example of the marriage of aesthetics to modular co-ordination'.[46] Alan Diprose concluded in the *Modular Quarterly* that, 'the Modular Society Pavilion at I.B.S.A.C. 1964 demonstrated that the theory of modular co-ordination is proved beyond doubt in practical application'.[47]

The third and final Assembly was designed by Bruce Martin for the 1966 IBSAC Exhibition using the metric 100 mm module (Figure 6.7) The structural framework was the A75 timber system produced by A.H. Anderson and designed by three members of Farmer and Dark Architects in the 1950s. Bruce Martin's design allowed for the names of the manufacturer to appear on all the components they contributed, so that the whole structure acted as a two-storey advertising stand for the participating firms.[48]

114 *The early years of modular co-ordination*

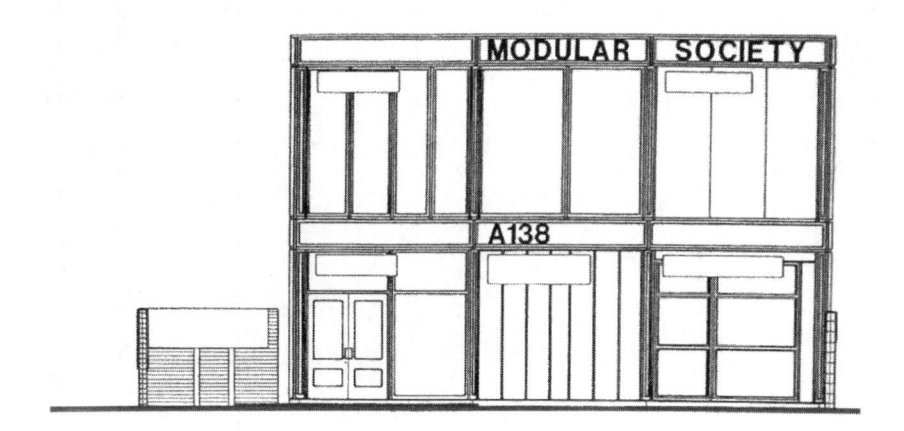

South-east elevation.

The Third Modular Assembly, 1966.

Figure 6.7 The third Modular Assembly at IBSAC 1966
Source: *MQ*, 1966.

The reviewer of IBSAC 1966 for *Set Square* noted that there were far too many gaping cracks on view between the cladding panels on many of the stands, giving the impression of hasty assembly to many of the pavilions, and not helping to promote the cause of industrialised building.[49] Diana Rowntree, writing for the *AD*, was impatient of the fragmented nature of industrialisation in Britain without any standardised dimensions or unified aims. She concluded that 'all our talk of new methods conceals the fact that we are still using industrial techniques to create a craft-type product, and with a craft-type labour organisation.'[50] By 1968, the tongue-in-cheek comments of the editor of *AD*, Monica Pidgeon, expressed growing cynicism with the whole field of industrialised building:

> Component systems for industrialised building seldom go further than a good joint, a universal cladding panel or a well worked out planning grid. Like nesting tables they seem to have boundless possibilities but their usefulness is strictly for the special occasion.[51]

The Modular Society did not produce any further practical examples of its work but continued to field members to sit as representatives on BSI committees (42 members in 1967) and campaign and educate the industry through its publications and public lectures (Figure 6.8). Despite the recorded doubts of some its members, both contractors and architects, the Society did not publicly address any of the problems in the site assembly of components revealed through its own experimental buildings.

Figure 6.8 Modular Society members at the Society's conference, 18 November 1963. From left to right: Alan Diprose (Modular Primer), P.H. Dunstone (Combinations), Mark Hartland Thomas (Secretary), Peter Trench (Chair), Bruce Martin (International Work)

Source: *MQ*, Winter 1963–64.

7 The BRS and the mathematisation of architectural modularity

> Modular Co-ordination is the key to the industrialisation of building. However, experience already gained in various countries has shown that the full benefit of modular co-ordination can only be achieved by international co-operation.
>
> EPA 174, First Report, 1956[1]

The aesthetics of number theory have been covered comprehensively by Lionel March and Philip Steadman (1971) where the authors make a distinction between modular co-ordination and proportional systems, noting that the former is concerned with the ratios of the lengths of components in any of the three directions found in a grid whereas, in the latter case, it is the proportional relationships between components that are critical. They go on to state that:

> This concern with lengths and ratios of lengths, with linear relationships, distinguishes modular co-ordination from many of the historical number systems employed in architectural design, where the greater interest tends to be in the proportions of the shapes which the system produces. Some ... proportional systems have incidental properties useful in modular co-ordination, and vice versa.[2]

Despite modular theory being presented as a practical, operational system for increasing the efficiency of building, Mark Hartland Thomas was also careful to advocate its use to architects using the language of aesthetics. Throughout the 1950s, proportion was a word used regularly in the presentation of theories of modular co-ordination by both the BRS and the Modular Society. Aware of the aesthetic preoccupations of their fellow professionals, researchers were careful to demonstrate that any new approach did not preclude the ability to harmonise proportion. Hartland Thomas had long held an interest in architectural proportion dating from his student days and did not consider that modular co-ordination, used as constructional system, was incompatible with architectural aesthetics. He considered geometrical proportion the 'shorthand notes of [cumulative] design experience, found by architects to be effective vehicles for the communication of fantasy'.[3]

There were a number of publications in the inter-war years that examined aesthetics and proportion through the application of mathematical analysis. Jay Hambidge, Yale Professor of Architecture, published *Dynamic Symmetry: The Greek Vase* in 1920 and in 1931 Matila Ghyka's two-volume treatise on the 'golden number' was published in Paris, with an introduction by the poet, Paul Valéry.[4] Ghyka's review of the historical use of the φ rectangle was published in English after the war in a shortened version as, *The Geometry of Art and Life* in 1946, and was cited by Colin Rowe in his seminal article of 1947, 'The Mathematics of the Ideal Villa: Palladio and Corbusier Compared'.[5]

Soon after the Modular Society was founded in 1953, the BRS announced the start of its own research into dimensional co-ordination and called on the Modular Society not to publish anything on module size before its studies were complete.

In a paper delivered to a Modular Society meeting in March 1955, William Allen presented the initial findings of the BRS research. His introduction presented the methodology of the study as a 'straightforward research and development programme'. He emphasised the scientific method with which:

> we are trying to reach what we believe to be the main objective of modular co-ordination, which is to have a simplified pattern of dimensions around which the whole of the industry, wet and dry, architectural and constructional, can organise itself.[6]

Espousing the cautious approach of a government-employed civil servant, he considered that even if the research took two or three years, it would be worth waiting for.

Allen's presentation contained a number of points which became contentious issues in the upcoming feud between the two organisations over modular co-ordination. After reviewing British modular work, which existed almost entirely in the field of prefabrication, he then explained that:

> studies of prefabrication alone obviously could give us only limited help in applying modular ideas to conventional building, for they omit the brick and block construction which is and will remain for some time the main substance of the industry.[7]

This deficiency had been remedied by a study of current German practice and Allen took care to point out that post-war German house building was largely traditional, using bricks and blocks, but with a requirement for modular planning and design where building was subsidised by the state. Dimensional co-ordination was standardised in the Federal Standard DIN

118 *Mathematisation of architectural modularity*

4172 which carried a number pattern for preferred dimensions for modular co-ordination in the form of a table. As Allen explained it:

> 'Preferred Building Numbers' . . . has two sides, one generating from a 12.5 cm unit and the other from 10 cm . . . The 12.5 cm side is used for bricks, blocks and all carcassing (i.e. constructional dimensions) while the latter is used for planning, each cell of a plan usually having its own 10 cm grid.[8]

The BRS had made a best practice study of the building of low-cost housing in Schleswig-Holstein, in which Allen explained that:

> Both freedom of design and reduction of cost are shown to be possible by the use of products whose dimensions are chosen from a series of related numbers . . . One cannot escape seeing a marked similarity between this development and some of the proposals of the Bailey Committee, and their implementation may deserve more effort than they received.[9]

After referring to Le Corbusier's *Modulor* as providing only a narrow range of 'aesthetic possibilities' for architects, the ground had been prepared for the introduction of Ezra Ehrenkrantz's work on preferred dimensions known as 'the number pattern'. Allen explained an outline of the idea and Ehrenkrantz elaborated on his theoretical work at a blackboard during the question-and-answer session. There were those present who freely admitted to not understanding the abstract mathematics behind Ehrenkrantz's work. Peter Gardiner and P.A. Denison of Cape Asbestos were both mystified and David Medd joked that the beauty of the 'number pattern' was that it allowed him 'to proceed exactly as he had been doing in the past'. Nevertheless, a number of Modular Society members, including Guy Oddie and Roger Walters, were impressed. Allen closed his talk by stating that the BRS's current work was to 'find a hypothesis' as a basis for a dimensional framework for modular co-ordination and then to test it. This was, to some, an inflammatory remark as the Modular Society had already come out, in 1954, in favour of the 4-inch module and many members were impatient to promote its use throughout industry.

The Modular number pattern

Ezra Ehrenkrantz, a 23-year-old MIT architecture graduate on a Fulbright Scholarship at the BRS, worked under William Allen on modular co-ordination for two years. A detailed account of his 'number pattern' was published in the Winter 1955 edition of the *Modular Quarterly*, in advance of the public meeting of the Modular Society where Ehrenkrantz gave a full explanation of his work, so that the audience were well prepared for the ensuing discussion.[10]

Ehrenkrantz positioned his work within the tradition of architectural proportional systems but attuned to the needs of mass production. Asking the question, 'Can a new scale be developed for our industrial age which will be able to include those unique properties and visual effects of the various proportioning systems which have previously been used in architecture?'[11] his answer was an unequivocal yes. The sizes generated from his proposed 'number pattern' made possible the use of whole number ratios found in harmonic proportion as well as the ratios necessary for dynamic proportion using the golden mean.

The overall aims of the BRS research were identical to those of the Modular Society, and the BSI: the rationalisation of the sizes of building components into a series of related components, which would result in cheaper building. Ehrenkrantz listed a number of associated conditions for this rationalisation. That it should be aesthetically neutral to allow freedom in design, take into account the properties of different materials, allow for the use of different proportional systems, be numerically simple and easy to operate, be applicable to both imperial and metric systems of measurement and take into account anthropometric data, optimal structural sizes and established planning sizes.[12]

His method was derived using nominal sizes on the assumption that 'building products meet at all joints and fill all spaces with standard parts'.[13] This completely side-stepped the problem of tolerances of fit that dominated so much of the debate within the Modular Society. Ehrenkrantz boldly and simply meshed together three bases of architectural proportion to generate a series of numbers, which could be used as standard sizes for building components. He took as his starting point the two series that make up the platonic lambda, the foundation of harmonic proportion, i.e. the doubling series: 1, 2, 4, 8 . . . , and the tripling series: 1, 3, 9, 27 . . . By ingeniously inserting a Fibonacci series into a graphical representation of the platonic lambda, he enabled the number pattern to include the possibility of proportions based on the golden mean of 1:1.618.[14] This three-dimensional framework was the means of generating what Ehrenkrantz termed 'The Modular Number Pattern'[15] (see Figure 7.1).

Ehrenkrantz argued that his system fulfilled all the criteria for successful rationalisation and standardisation of building components. However, his dismissal of the use of a single base module and his defence of the British brick opened up his proposal for attack from Modular Society members:

> It may be possible to determine a single base module for a particular group of products, but to try to do so for the entire industry at this stage can and ought to be questioned. For the present, it would seem impossible to disregard the solid clay brick which has an optimum nominal size of 3 inches × 4½ inches × 9 inches.[16]

120 *Mathematisation of architectural modularity*

1	2	4	8	16	32
2	4	8	16	32	64
3	6	12	24	48	96
5	10	20	40	80	160
8	16	32	64	128	256
13	26	52	104		416
21					—

BRS MODULAR NUMBER PATTERN

3	6	12	24	48	96
6	12	24	48	96	192
9	18	36	72	144	288
15	30	60	120	240	480
24	48	96	192	384	—
39	78	156	312		
63	126	252			

BRS MODULAR NUMBER PATTERN

9	18	36	72	144	288
18	36	72	144	288	—
27	54		6	432	
45	90	18	0		
72	144	288	576		
117	234	468	—	—	—
189	378	—	—	—	—

BRS MODULAR NUMBER PATTERN

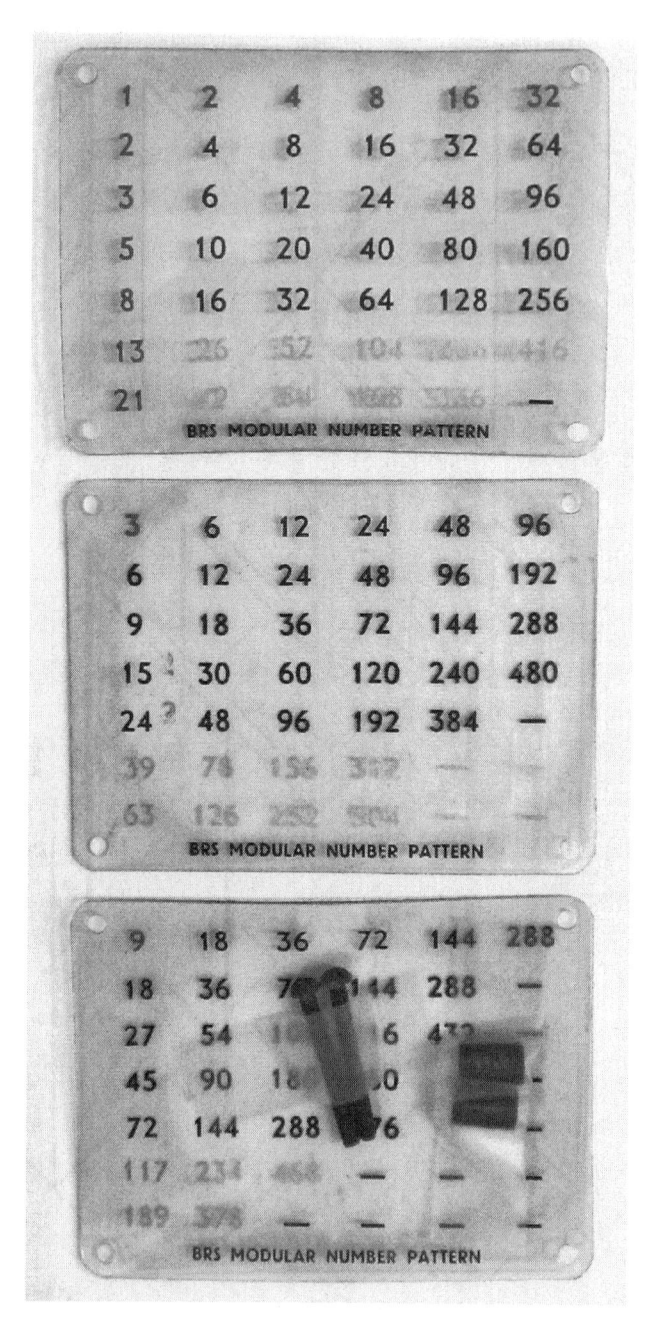

Figure 7.1 Pieces of a three-dimensional model of Ehrenkrantz's number pattern found inside the back cover of *The Modular Number Pattern* (1956)

Source: Cambridge University Department of Architecture Library.

This set Ehrenkrantz, and effectively the BRS as well, at odds with the work of the Modular Society. Ostensibly Ehrenkrantz had exactly the same aims as the Modular Society architects in seeking to find an architectural solution to the problems of industrialisation by ensuring aesthetic flexibility within a standard range of products. However, his approach, and that of the BRS, were careful not to antagonise the brick industry and to be inclusive of traditional methods of construction. He concluded that his number pattern 'makes the whole of the building industry a closed system, where all products combine with each other'.[17]

Mark Hartland Thomas, for the Technical Committee, drafted the Modular Society's formal reply to Ehrenkrantz's presentation of his work:

> To summarise our views upon Mr Ehrenkrantz's paper and the discussion on it, we would say that we accept his Number Pattern as a useful piece of apparatus, whilst regretting the advocacy of some who claim it as a complete solution to the problem of dimensional co-ordination of building components. We say that the pattern should be given scale by using his numbers as multipliers of the four-inch module. Finally, we must express regret that so much time has been spent in producing a piece of apparatus to demonstrate the well-known requirements that preferred sizes of components must be such as to combine together in building construction, whilst the actual selection of such sizes is delayed and the attack upon the key problems of Thickness and Jointing, and the establishment of conventions for small-change dimensions, has hardly begun.
>
> 9.2.56[18]

On Ehrenkrantz's return to the United States, he developed the very successful School Construction System Development programme in California which was exported throughout the USA and also used in Canada.[19] Ironically, for this he abandoned modular co-ordination in favour of a 5-feet planning module chosen because it could accommodate 4-feet fluorescent lighting tubes.[20] Although Ehrenkrantz's aims were co-terminous with the Modular Society's own, his solution was rejected.

EPA Project 174

In 1953, Bruce Martin became Head of Modular Co-ordination Studies at the BSI as well as the British co-ordinator of the European Productivity Agency (EPA) Project 174. The EPA was an organisation created in Paris ostensibly as an offshoot of the OEEC, and funded almost entirely by American dollars through the Marshall Plan.[21] Inaugurated to promote productivity 'and thereby raise European standards of living', its primary aim was 'to convince management and workers alike of the benefits of productivity and to enlist their co-operation', aims that aligned with those of the earlier AACP (see Chapter 4).[22] The Advisory Board comprised prominent figures from European industry and organised labour. Between 1953 and 1960, when the EPA ceased functioning, a generous budget had financed a vast amount of research, training and education programmes across all industry sectors, with the overriding aim of introducing American management practices to European industries.

EPA Project 174 consisted of collaborative research between 11 countries into modular co-ordination as a method to increase productivity in construction. [23] Phase I comprised a review of existing approaches by each participating country in order to reach common ground for formulating modular theory on an international basis. For the UK, Phase I of the project involved the BSI team, under Bruce Martin, investigating the dimensions of building products while the Building Research Station examined 'experience in the field' under the direction of William Allen.

At the end of Phase I of the EPA Project, a comprehensive policy statement was agreed by the participating countries stating that they should, 'work towards the establishment of a modular system having as far as possible a similar character, a single Basic Module and similar sizes in each country'.[24] The size of the basic module was agreed as 10 cm in the metric system, and 4 inches in those countries using the feet/inches system. An exception was made for Germany where the 12.5 cm module for carcassing work was already in use alongside the 10 cm module for finishing work. The majority of participating countries favoured the 4 inch or 10 cm module and by 1955 five of the 11 participating countries (Belgium, France, Italy, Norway and Sweden) had already standardised the basic module at 10 cm.

The problem of integrating brickwork into an international modular system was tackled early in the project. There were two possible solutions: either the wide range of brick sizes found in the participating countries should be re-dimensioned to fit with a 4 inch or 10 cm module, or brickwork should be designed to overall modular dimensions using existing brick sizes. Not surprisingly, the latter was chosen as a working solution although the final report did state that modular bricks were preferable and called on manufacturers to modify their products over time.[25]

Brick sizes varied greatly between countries and also within countries. Italy produced five different sizes, France two, Denmark three, Greece five, Belgium three, Sweden three, and the UK two. The Netherlands and Norway produced only one size.[26] None of these brick sizes fitted with the proposed international modules but as the majority of the participating countries did not attempt to derive a modular system based on the size of their domestic bricks, this did not affect the development of modular theory. However, representatives from the BRS still argued against the implementation of a 4-inch module on the grounds that the majority of British building was in brick. Some 61 per cent of British brick production was in pressed brick, necessitating the use of moulds in manufacture that were difficult and expensive to alter. They therefore insisted that any system of modular co-ordination must include a 3-inch module that related directly to brick size, the brick being the most important component in the British building industry. (The Netherlands relied on 73 per cent of pressed bricks and also fielded a powerful brick lobby.)

The British brick industry had reacted to the threat it perceived from the prefabrication movement by promoting and advertising its products widely. In 1947, the National Federation of Clay Industries launched *The Brick Bulletin*, a journal devoted to fostering 'a fuller understanding of the nature and the best use of the brick', allying the use of brick with the best of English building.[27] This publication made public the brick lobby's argument:

> The standard brick is in fact a building 'module' – to use one of the ugly but concise semi-scientific expressions now current – a unit 9 inches × 3 inches (including the joint – we are now referring to the smaller 8¾ inches × 2⅝ inches brick) is the useful function of its design.
>
> Present brick sizes did not 'just happen', nor are they imposed by the whim of the brickmaker. They have been developed and adjusted through long usage, and are an important factor in building design and construction.[28]

The membership of the Modular Society, despite its pronouncements of being representative of all the participants in the building process, not only lacked representatives from the operative workforce, overlooked to the extent of both invisibility and dispensability by many members, but, more importantly in relation to their aim to institute a 4-inch module, included only one brick manufacturer.

Arguments and reassurances from the Modular Society and Bruce Martin at the BSI that modular theory presented no inherent threat to the brick industry were ignored by the BRS. In the same year, 1956, that Ehrenkrantz published his number pattern, the BRS revealed that their test buildings, to be built as part of the second phase of the EPA Project 174, would be in rationalised traditional form and not based on the 4-inch module. From Bruce Martin's perspective, as the British co-ordinator of EPA Project 174, the BRS proved to be a difficult research partner, working in secret and withholding their findings from the BSI, and a national embarrassment in their flouting of the agreed methodology of the EPA Project. Alfred Bossom had even tabled a question to Parliament asking:

> Why the BRS is now attempting to discourage the adoption of any module and is intending to design the UK test buildings in the EPA without a module, in spite of their agreement with the other EPA countries of August last?

124 *Mathematisation of architectural modularity*

The Parliamentary Secretary to the Minister of Works replied:

> I am informed that the BRS is not attempting to discourage the adoption of any module but is reserving its own decision until the experimental work now in progress is complete. In designing the buildings to which the Honourable Member refers, the BRS is applying knowledge of the use of modules and co-ordinated dimensions without confining itself to the use of any particular module.
>
> 26 April 1956[29]

In June 1956, Bruce Martin, now a member of the Modular Society and its Technical Committee, suggested that William Allen be invited to join the Modular Society in an effort to increase communication between the two research bodies. Allen deferred to his superior, the Director of BRS, at that time Dr F.M. Lea, and according to Martin, in tune with the brick industry's interests. Approval was granted and the minutes of the Modular Society Technical Committee for 2 October 1956 show William Allen was present, although no immediate rapport between the two bodies emerged.[30] Finally, in May 1957, the Council of the Modular Society held a press conference at the Royal Society of Arts in which Mark Hartland Thomas, acknowledging the lack of agreement on module size, announced an adjustment of policy to take into account the British brick.

> At present about half the building industry uses brickwork as its prime material, and most of it is in housing, flats or building of a similar scale. The common British brick is not related in any simple way to a 4-inch module and while this does not present serious difficulties on large buildings, it does to a greater extent on small ones . . . The most practical course to adopt for this sector of the industry seems to be to accept the brick size much as it is for the present and to continue to relate the size of components needed for this particular work to the size of the brick.
>
> . . .
>
> The Society naturally regrets that it does not seem practical to press the general policy of a 4-inch module uniformly throughout the British building industry at present. They believe, however, that the present recommendations offer the best prospect of the widespread use of the principles of dimensional co-ordination in the immediate future, with a minimum of uncertainty and disturbance to the existing industry and standards; and they look forward to a future in which one modular system alone prevails over the whole field of building.[31]

On the platform with Hartland Thomas for the occasion of the press conference were William Allen (BRS) and Bruce Martin and P. Cutbush,

Mathematisation of architectural modularity 125

representing the BSI as well as Donald Fraser of Unit Construction, G. Laurence of Sussex and Dorking United Brick Company, David Carter of Carter and Company and the structural engineer R.A. Sefton Jenkins. William Allen applauded the efforts of the Modular Society but emphasised the ongoing and experimental nature of BRS work on dimensional co-ordination. Bruce Martin spoke at length on the European and global context for agreeing an international module as the basis for standardisation of building components. He reiterated his argument that no common unit of size for brick existed and that the concept of modular brickwork was far more useful.[32]

Lord Bossom arrived late and contributed 'a few words' on his experiences of the 'rapid speed' of building in the USA in 1909 based on simplicity of construction and the willingness of the bricklayers to use any sized brick given them. As the discussion drew to a close, there appeared to be some agreement emerging between those on the platform, if not on module size, at least in relation to a common viewpoint of the building industry as a 'very conservative industry'. This was summed up by Bruce Martin's description of the 'unbreakable circle' that prevented innovation: 'the manufacturers will not make the parts until they are asked for, and the architects will not ask for them until they are made'.[33]

Phase II of the EPA Project demonstrated the two very different approaches being taken on modular co-ordination in the UK. Whereas the BSI submitted as their experimental building the fully modular single story administration and laboratory blocks designed by Bruce Martin at Hemel Hempstead, the BRS submitted a report on the building of two versions of a traditional brickwork terrace, one in load-bearing brickwork and the other using load-bearing cross-walls with prefabricated panels for front and rear façades (Figures 7.2–7.5).

The last EPA meeting was held in London in April 1960 when the members continued their association as the International Module Group and agreed to publish their transactions as a supplement in the *Modular Quarterly*. The final EPA Project 174 Report was published in 1961 and printed the views of BRS as footnote explaining their individual stance on brickwork together with an Appendix setting out in full the number pattern theory devised by Ehrenkrantz.[34] At the close of the project Bruce Martin and Ernst Skarum of the Danish Building Institute, collaborated on a short promotional film on modular co-ordination. As described by Martin, the last scene is a whimsical fantasy sequence showing numerous building components sprouting wings and flying around the globe until they all settle into their appropriate modular spaces, presumably on modular grids.

The work of the BRS and BSI, and their oppositional theoretical stances on modular co-ordination, continued beyond the EPA-funded project and contributed to the formulation of Part II of BS 2900. When the BSI Study Team produced, in 1959, a Draft for Part II, *The Module and the Range of Modular Sizes for Building Components*, which specified a 4-inch module,

126 *Mathematisation of architectural modularity*

UNITED KINGDOM I.
Hemel Hempstead:

Offices and laboratories.

Figure 7.2 Photo of British Standards Institute test buildings, EPA Project, designed by Bruce Martin

Source: European Productivity Agency (1961) *Modular Co-ordination, Second Report of EPA Project 174*. Paris: OEEC.

the BRS responded by producing their own alternative draft document.[35] In this they continued to refuse to accept the agreed international module size of 10 cm on the grounds that it was incompatible with the 3-inch height of British bricks. Their position caused much frustration to Bruce Martin at the BSI and also to the aims of the Modular Society as extracts from the following letter reveal. Marked 'Private and Confidential', it is addressed to the BSI Technical Committee charged with producing the British standard on modular co-ordination and is a criticism of the BRS document *Alternative Draft Proposals for Part II of BS 2900*:

> In questions of standardisation the presumption is in favour of conformity. A nonconforming scheme of standardisation is almost a contradiction in terms, but that is what the BRS has proposed. To win acceptance, such a proposal must offer very substantial advantages in order to compensate for the obvious disadvantages of nonconformity.
>
> . . .
>
> The BRS proposal which the Station has taken seven years to produce, offers no such advantages over the international system. On the contrary we find it inferior on every count.[36]

UNITED KINGDOM I. Hemel Hempstead:

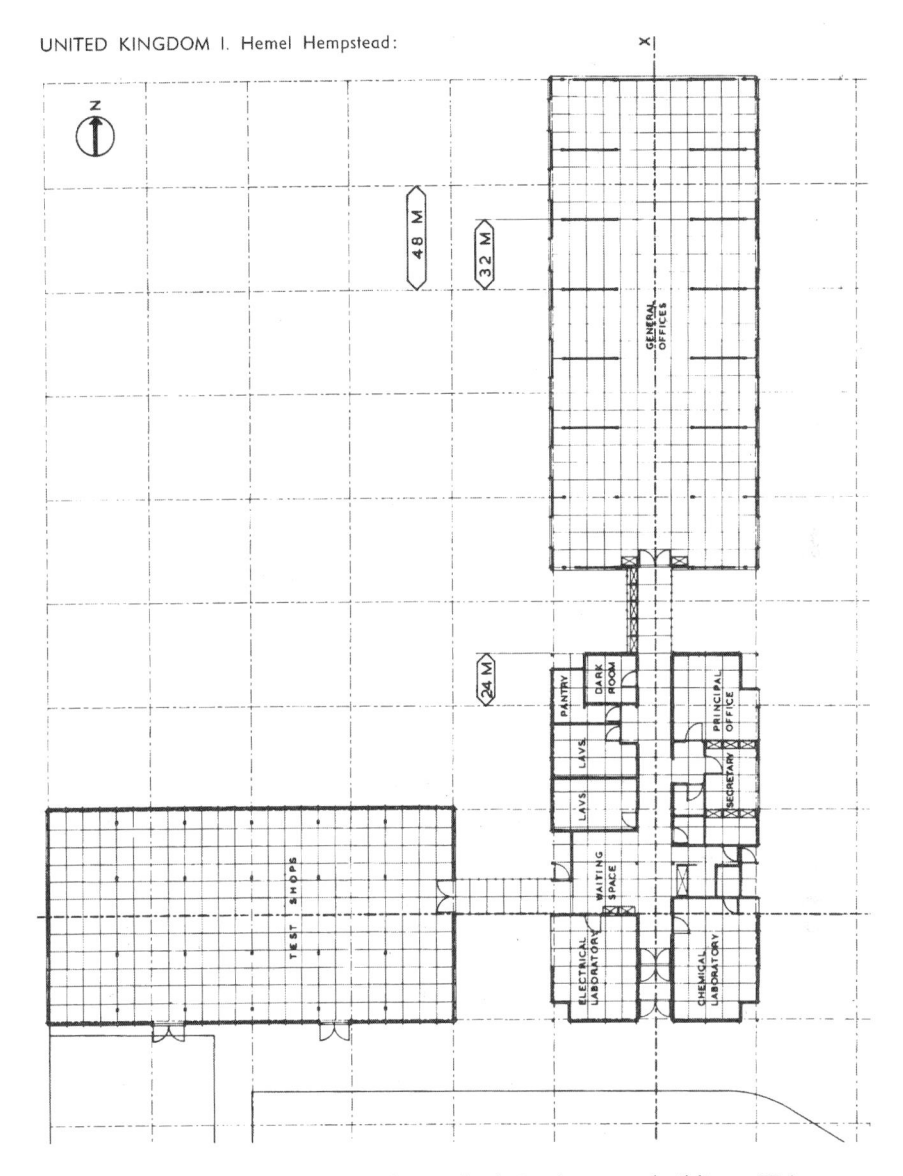

Figure 7.3 Scale drawing of British Standards Institute test buildings, EPA Project, designed by Bruce Martin. Plans based on four-inch module.

Source: European Productivity Agency (1961) *Modular Co-ordination, Second Report of EPA Project 174*. Paris: OEEC.

128 *Mathematisation of architectural modularity*

UNITED KING-
DOM II. Hatfield:

Terrace houses,
south elevation.

Figure 7.4 Building Research Station test buildings, EPA Project. Rationalised traditional terraced housing in brick and block.

Source: European Productivity Agency (1961) *Modular Co-ordination, Second Report of EPA Project 174.* Paris: OEEC.

The letter continues making four main points against the suitability of the BRS proposals on technical grounds, having already pointed out the economic disadvantages of Britain being the only country in the developed Western world not to ratify the 4-inch or 10-cm module.

1. *Correspondence with brick dimensions*
Despite the origin of the BRS proposal lying in the conviction that the current nominal sizes of bricks should dominate any system of dimensional co-ordination there is a poor correspondence with existing brick sizes . . . The reason being the arbitrary omission of many multiples of the basic unit from the BRS scheme of 'preferred dimensions'.

2. *Correlation of component sizes and building spaces*
The BRS scheme is essentially one of gaps. Any system that arbitrarily prefers some dimensions and rejects others is bound to lead to unpredictable design sizes, which do not correspond with the preferred range of components, but progressively score more and more misses as the sizes get larger.

In the EPA system any combination of modular sizes is equal to another modular size and a modular space. Correlation is perfect.

Mathematisation of architectural modularity 129

UNITED KINGDOM II. Hatfield:

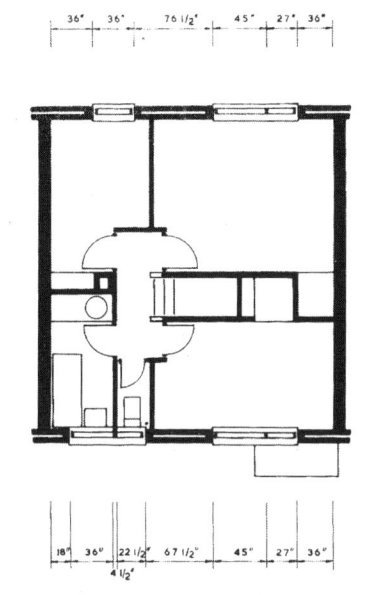

Terrace I: Plans.

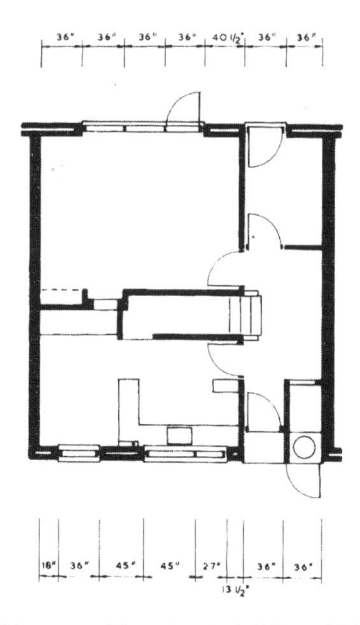

Figure 7.5 Plans of BRS terraced housing in brick and block
Source: European Productivity Agency (1961) *Modular Co-ordination, Second Report of EPA Project 174*. Paris: OEEC.

130   *Mathematisation of architectural modularity*

> 3. *Correspondence with existing sizes*
> The extraordinary anomaly of 'occasional ad hoc departures' which involve the casual importation of dimensions not in the original preferred range ... Again, this does not compare well with the EPA system.
>
> 4. *Tolerances*
> The BRS proposal 'does not depend on cubical grids of any kind' but grid planning is not obligatory under the EPA system either.
>
> It is traditional practice for a builder to set up profile-boards and to strain lines between them in order to afford a reference system from which to control the dimensions of all parts of the building. He uses in fact a grid, since he takes this system of reference up ahead of construction into the third dimension, it is a 'cubical grid'.
>
> The BRS avoidance of 'cubical grids of any kind' is thus a negation of traditional practice.[37]

In 1961, George Atkinson succeeded William Allen to direct research on modular co-ordination and the BRS produced a draft document written by Atkinson that did not include Ehrenkrantz's work on preferred sizes and predicated co-ordination on the 4-inch/10-cm module. Hartland Thomas wrote to Bruce Martin arranging a meeting prior to the upcoming BSI Technical Committee meeting in order to 'concert our attitude' towards the BRS proposals.[38] Nothing new transpired, however, and the same arguments were being played out a year later, to the great frustration of Mark Hartland Thomas, now acting as an international expert on modular co-ordination for the United Nations. In 1962, he was still eliciting support from Bruce Martin in what, judging by the wording of his correspondence, was becoming almost a personal crusade against the BRS and their entrenched antagonism to the 4-inch module.

> About B 94/3, the thing for you to watch with the eye of a hawk is lest BRS slip in some dirty little crack against the 4-inch module in their explanation of their table and how to use it. If they do manage to explain their cockeyed ideas entirely without false accusations against modular, it will be the first time ever.[39]

In the same year, 1962, the RIBA Council finally pronounced on the module, coming out in favour of 4 inches, 'as an interim measure' while the BSI Committee continued their deliberations.[40] This provoked Mark Hartland Thomas into defensive action on behalf of the Modular Society, eager to preserve the society's reputation and role as the organisation in the forefront of campaigning for dimensional co-ordination (see Appendix 2).

However, opposition from the BRS against the 4-inch module and the ensuing stalemate at the BSI continued for another four years until a truce, of sorts, came about with the issue of BS 4011:1966, which used metric measures.

Conclusion

It is difficult to know exactly why the BRS refused to ratify the 10-cm or 4-inch module for so long. While Modular Society members suspected that lobbying by the brick industry was behind the long impasse, it will take further research on building materials during this period before any clear conclusions can be reached. Industrialisation continued without it anyway.

Peter Goodacre reviewed the era of dimensional co-ordination in 1981 for the SRC-funded research project instigated by the Modular Society. The conclusions were that:

> [U]ntil major changes of attitude are made, there is little that dimensional co-ordination can do to significantly affect the economy of the construction industry. It was perhaps hoped by some that dimensional co-ordination would facilitate the continuing split between design and production by providing a readily available set of parts like a Lego or Meccano set which would always be capable of fitting together. This has not proved the case and it thus seems that there must be a closer integration between design and construction.[41]

Part III

'Never argue with the architect'

Architects and building workers 1940–70

8 'Put nobody between the architect and the men'

The role of architects on site

> We got criticism from architects who said we were too obsessed with process and really it was the product that counts, and that when we were all dead and gone, it was the buildings that remained that would really count and nobody was really interested in how they were produced.
>
> Alan Meikle[1]

Introduction

In the post-war years, Nottinghamshire, like the rest of the country, had an acute shortage of schools but with the added problem of subsidence, caused by defunct mine workings, at many of the sites on which they were to be constructed. When Donald Gibson left Coventry to become County Architect for Nottinghamshire in 1955, he had, by the end of the year, attracted a number of architects from Hertfordshire to join him. Henry Swain became head of the Development Group; Dan Lacey became assistant county architect and Alan Meikle and a number of others from the school building programme took up posts. They became core members of the team that devised the spring-loaded Brockhouse steel frame able to withstand the movement effects of subsidence. This frame was manufactured locally and became a central component of the CLASP building system.[2]

The Hertfordshire architects had become well known for their innovative 'rolling programme' consisting of design, production, feedback and development in their efforts to produce new schools under the tight restrictions of post-war shortages. The same approach was applied very successfully to school building in Nottinghamshire and by the late 1960s building in CLASP had extended from schools to include police stations, old people's homes and other examples of local authority architecture. Efficiency limits had been reached in the off-site manufacture of CLASP components and, despite much of the school building programme being serially contracted to the local firm Searsons, by the late 1960s, Henry Swain decided that the record on site assembly did not compare well with factory production. In 1967, he set up the Research into Site Management (RSM) Project expressly to 'contribute to the further development of the CLASP prefabricated system by focusing the attention of the designers more sharply on the problems of site assembly'.[3] This was to be achieved by the architects becoming directly involved in the process of building by being based on site and directly employing the men who were going to assemble the building. The proposal was a replication of Lethaby's ideas on decreasing the distance between design and production through the actual presence of the architect on site and the hope of an ensuing improvement in communication with the building workers. The RSM project was, however, not the equivalent of setting up a mini DLO as the number of actual buildings had to remain a small proportion (7.5 per cent) of the County building programme for political reasons. Initially funding had to be applied for on an annual basis in order to appease both the Conservative councillors with their commitment to the open market and the local contractors and their antagonism to the DLO. The ethos was very different to that found in either a DLO or a general contractor in that design and construction were to be united in one organisation.

The core function of the RSM project was to bypass the contractor entirely and for the architects to become directly involved with the difficulties of site organisation and assembly. It was an opportunity for the architect to become fully conversant with the entire process of construction. Alan Meikle, who directed the project, elaborated on the context for initiating this particular direction of research:

> We had made huge advances in manufacturing components in buildings but the missing bit was how all this was put together on site . . . When you went onto the site, what you found were people knee deep in mud, and making a terrible mess of things and doing things in a totally disorganised way. It probably wasn't as disorganised as it looked, mind you . . . but clearly there were a lot of inefficiencies, there was an awful lot of waste. Some studies said that 10 per cent of all building materials were wasted on site . . .[4]

But Meikle also had personal reasons for being very interested in the project. Not all of his experiences of modular building at Hertfordshire had been positive and he recounted an incident he witnessed when a vast piece of glass, destined for a gymnasium window, was delivered to site. When Meikle questioned the foreman as to why it was not in two pieces to fit above and below the transom, to match the other windows, he replied that there was nothing shown on the drawing and that they had been told that on no account were they to question anything that was on the drawings. It was, of course, an error on the part of one of the architects and illustrates the response of site workers to an authoritarian management structure. Meikle summed up his experience:

> I thought that actually typifies the problems that arise when the architect says, 'I am the man who makes the decisions, you must never ever ques-

tion them', even if they appear to you, like to the foreman . . . something silly.[5]

Meikle was also, prior to the setting up of RSM, Chairman of the Barrier Group, an informal discussion group of around 50 construction professionals who met occasionally in London. John Carter, the architectural journalist, was the secretary for a time. It was during these meetings that Meikle noticed that the architect members were, on the whole, distinctly uninterested in what happened on site. As he saw it, 'There wasn't a proper connection between the designer's brain and what was happening at the end of his hand.' Meikle attended a Henley Management Course where he was one of only two architects on a course of 66 people. He returned inspired by the potential of management to change procedures on site, to the extent that, 'Henry [Swain] felt I was getting a bit big for my boots. He felt I needed something to apply the lessons I had learnt.'

One of the radical departures from orthodox practice that gave the Hertfordshire school building programme the deserving retrospective description of 'social architecture' was the incorporation of user group studies into the design process. The continuous development programme relied on feedback from teachers and on research into the changing pedagogic practices of post-war state education. The RSM project, although justified by Henry Swain using the terminology of costs and efficiency, can perhaps better be described as an attempt to use human relations management in construction. From the early days of prefabrication, commentators had often stated that industrialisation would provide benefits to the workforce, although these had been envisaged mainly as the removal of large numbers of workers from the site and into the more comfortable and stable conditions of the factory. It was extremely rare for the exponents of industrialised building to consider that the operative workforce had anything to contribute to technological improvements, and it is questionable as to how far site operatives' knowledge was incorporated into the feedback loop of continuous development even though it appeared, as a dotted line, on the project operational plan. There was never any scope for giving management responsibilities to workers so that input was always indirect via the white-collar staff and, as one job architect with seven years experience of working on CLASP designs put it, 'by keeping operatives out of management teams, we are wasting one of our most valuable resources'.[6]

Discussions on the setting up of RSM started in 1966. The project started fully in 1967 and used the CLASP Mark 4B system for the first three years of work and then tested the Mark 5 in 1969, building mainly primary schools and old people's homes before the unit was wound up in 1983. In 1971, Henry Swain wrote an appraisal of the first few years,[7] focusing on management and organisation, in which he was keen to avoid describing the results as a model for improvement for the industry. Instead he referred back to Lethaby, introducing the experiment as, 'a small contribution to the large and difficult art of building'.

Every RSM job was divided clearly into three stages: design, construction planning and construction. The RSM personnel consisted of the architects,[8] an architectural technician, responsible for costs, a quantity surveyor responsible for contracts, a materials manager, two accounts clerks – one for payments, one for costs – and the core operative workforce consisting of nine carpenters, one bricklayer, and six labourers. Subcontractors were used for plumbing, felt roofing and other small parts of the scheme.

The following section describes the activities that were unique to RSM, for professionals and operatives, under the three stages of design, planning and construction.[9] This account refers to the building of the first eight schemes of the project, all schools except for one old people's home.[10]

The architect responsible for an RSM job had to determine the finished product and also take charge of the process of bringing it into existence. This involved directing a 'medium-sized industrial enterprise', being responsible for all decisions taken on site and giving detailed instructions for each part of the work to the operatives. All this required carefully thought out and practical designs, or 'buildability' to use the jargon of a later era. It also harked back to earlier calls for the architect to resume control of the building process, from Lethaby at the start of the century to later writers in the immediate post-war years worried at the encroachment of other professionals onto the architect's turf with increasing industrialisation. Using the graphical representation of contractual relationships formulated by Higgin and Jessup in the Tavistock studies, Swain depicted the relationships found in the RSM project as closer to those of the eighteenth century than the twentieth.[11]

The economic necessity of keeping a small group of operatives continuously employed was the factor that drove the speed at which the office staff worked. The Programme of Work for the architects, surveyors and office-based staff was drawn up working backwards from the site operations shown on the strategic programme. These activities were costed and checked against the total cost sum for the entire job and at this stage any modifications to the design that might affect the building process were made.

Costing procedures and the wage system

From the outset the RSM project had to be completely transparent in its accounting and costing procedures, partly because it had to regularly prove to Conservative councillors that the experiment was not being subsidised by the local authority, thus giving results which could not be comparable with small local firms. The project, in order to refute any accusation of subsidy, allowed for 155 per cent overheads on each job to include rent of the office in County Hall, cleaning, printing, computers etc., considerably more than most medium-sized contractors. But it was in finding the actual and real cost of each element of the building that became the factor that was crucial to the success of RSM.

Although it was known that the 'real' cost of building was the sum of all the cheques paid out to operatives, contractors and suppliers, the project found that they needed to have this information, accurately, and on a weekly basis to prevent going over budget. Expenditure on wages in one week, complicated by different rates for the different trades, NI contributions, sickness payments, etc., did not bear a direct relation to the actual cost of labour in that week. Similarly, payments for materials and components did not correspond to the actual progress on the job. It transpired that the RSM team had attempted to cost at accuracy levels that were unrealistically high; the solution was a simple one – to accept 1 per cent inaccuracy and to use an average rate for labour costs.

The RSM quantity surveyors, Henry Morris and John Bennett, devised a cost control system for weekly recording of labour costs based on the man-hours spent on each operation or part of operation using data from the time sheets. The job architect assessed the percentage of the operation that was complete and this was compared with the sum of money allowed in the budget for the work completed in that week. The state of expenditure on materials was also estimated against the allowance in relation to the budget to give a weekly check on running expenditure. Once these two documents became a regular part of the Monday morning planning for the job architect, there were no major financial crises. Cost parameters were so tight on RSM projects that all the office staff were aware of the relative costs of professionals compared to building workers and, wherever possible, unnecessary professional time was cut.

Out of the RSM experiment came a costing method that referred to the amount of work involved in discrete operations, where the operations measured were called 'features'. It also dispensed entirely with the traditional bill of quantities. The 'features' were clearly defined and given a notation in the Building Industry Code devised by Morris and Bennett for the CLASP Development Group. Swain suspected that contractors' prices were arrived at in a much less precise fashion, however, he did not criticise the builders for this but rather the inaccurate and unwieldy method of pricing set out in the Standard Method of Measurement (SMM). It is not surprising, given the drastically different nature of the duties asked of the quantity surveyors compared with their usual professional practice that many were unable to adapt to RSM procedure and in the first few years 16 quantity surveyors passed through before a stable staff group was achieved.

Management strategies, productivity and wage systems were interconnected in the post-war years through the contentious issue of incentive and bonus payments. The wage system chosen by Swain for the RSM project was another factor that set them apart from the typical medium-sized contractor and it flew in the face of accepted private-sector practice in the construction industry.

> [W]e are not using bonus systems based on output. We think that bonuses encouraging greater production from the men make it harder

for management to ensure really good workmanship. The quality of a CLASP building to a large extent depends on accurate setting out, care of materials in store, careful handling and, above all, well organised sites. Without an output bonus system it has not been too difficult for us to achieve these.[12]

Swain would very probably have been aware of the arguments against bonus and incentive payments put forward by the construction unions. The decision was also influenced by Meikle's management training at Henley and his personal interest in the new, human relations approach to management emerging from the USA. He had read several books, in his words, 'called things like "Piece Work Abandoned"' and was aware of studies done in America which showed that piece work was the anathema of team-working. In conversation, Meikle revealed:

I wanted to move away from wage based to salaries but we never quite got that far. We did just about get to profit sharing . . .

I said, whatever happens, we are not going to have payment by individuals or payment by [gang] results. So that we didn't give any bonuses for chaps who might be working faster than others, but perhaps not quite so well. But we did have bonuses related to overall productivity at the end of the job . . . But the general principle of not paying them according to the number of bricks that they lay or the number of windows they'd glazed, that was established early on. They were on hourly rates with a bonus related to overall team performance.[13]

Very early in the project it was discovered that the architect's joinery details, supposedly drawn to provide information for the carpenters working on site, were being interpreted by the foreman verbally and by using quick sketches; the actual working drawing was left behind in the hut. The job architect thus learned to modify drawings on site to make them more relevant, and with time these became more precise.

The early schemes used a system of drawings called Detroit Sheets; each sheet was composed of a sketch of a specific CLASP assembly, cross-referenced to other drawings where necessary, together with materials and labour time for the task described on the sheet.[14] These sheets described the units of work that made up the whole job. They were also fundamental in pricing a job and provided the basis for costing on a *pro rata* system, instead of using the SMM method to produce a standard bill of quantities. They thus represented not just the end of the traditional method of working for a quantity surveyor but the beginning of a new way of communicating on site with the operatives. They also undermined traditional site processes by introducing the format and ideology of factory assembly to the building site. Operational planning was key to the success of RSM, especially as building with components proceeded at a faster speed than traditional building and many decisions had to be made quickly by the job architect. His presence on site throughout much of the construction period facilitated this. With the strategic plan in place, two weeks before work was due to begin on site 'tactical' details were programmed.[15]

With each gang leader in possession of a work sheet containing all the necessary information, management was based on agreed objectives, and, according to Swain, less site supervision was required. It also gave considerable responsibility to the men for planning their own work and explaining the reasons a job might be taking longer than expected. The system made it possible to trace whether hold-ups were due to oversights in planning, lack of motivation from individuals, or technical problems on site. The gang leader filled in the work sheet and any problems or changes to the usual procedure were recorded. Work sheets were analysed and the information they contained fed back to the job architect and used as a basis for planning the next job. They became the formal means of collecting information on site processes and acted as the operative part of the feedback loop.

Site communication

Communication became a familiar subject during the 1960s in debates on management. Most of the studies specific to the construction industry focused on the difficulties caused by the large number of professions involved. When 'the men' were included, they were usually referred to only in terms of effective labour resourcing and the use of payments and bonus schemes. When Henry Swain published his account of the project in the *Architects' Journal*, he made the unusual step of including the names of the operative, as well as the staff, members of the team.[16] This was a clear indication of his respect for the working man and the operative team that had stayed with the project for over two and half years. This was very unusual at a time of rapidly increasing self-employment and the very high wage packets that could be earned on the 'lump'. The stable group of operatives that the project attracted had a large part to do with its success. With each successive school built, output increased and the unit became highly efficient. The output per man (calculated by dividing the total project cost by the average number of men working) increased steadily with each project undertaken. At the end of the first eight schemes the unit had made £37,043 savings, an average of 5.89 per cent saving to the client on each building.

This stability of the workforce and their familiarity with the design of CLASP schools and the fit of the components enabled the use of less formal methods of communication between architects and operatives. As Alan Meikle recounted:

> Later on, of course, we made use of that ability to use shorthand because once we had got to the point where the men had been trained in our way

of doing things, we didn't have to give them drawings. Jumping several years right towards the end of this, I had a conversation with a carpenter about a recessed doorway we wanted in a school and I said to him, 'Now look it is going to take about ten days to draw this thing out. It is quite a complex piece of work but it will only take you a couple of days to do it and it seems to me ridiculous that it takes us ten days to draw something that takes two days to build.' So he said, 'Well, let's talk about it.' So I spent about an hour with him sitting down and doing some sketches and told him what the objective was and what the performance specification was and he listened very carefully and probably made one or two notes too. And at the end of it, he said, 'Well, what you are really saying, Mr Meikle, is that you want this recessed door to look like the school we did two years ago at Bingham but instead of using brickwork you are going to use concrete panels.' And I said, 'Yes, that's right.' And he said, 'And instead of that door which was solid at Bingham you are going to have a half glazed one like the one that we used at Cotgrave.' And I said, 'Yes.' And he said, 'And in addition to that we are going to have a tiled entrance and not an *in situ* one.' and I said, 'Yes.' And he nodded a bit and said, 'OK, I've got it, leave it with me. I'll do it.' And off he went and did it and no formal communication passed.

But it was a clear illustration that as the result of two or three years of close collaboration he knew my method of working and I knew enough about him to be able to trust him to do what I wanted. And of course the finished article was probably better than I could have drawn out because he had added one or two little practical points that you didn't appreciate, you know. You can't get a flashing in there like that and that sort of thing. So that clearly, that old communication had totally disappeared . . . [17]

The end of the 'old' method of communication had the added advantage of saving money on architect's time. With the establishment of mutual trust, information began to flow and the building workers became active and constructively critical of the process. It required time and strong examples of the good intentions of the staff, however, before this became a routine part of site life. One of the earliest demonstrations of the commitment of the staff to the rhetoric of the project is the incident below recounted by Alan Meikle and an unusual and very powerful example of an architect siding with the workforce against a manufacturer.

If there are mistakes made and there always are in building, there has been a tendency to pass it down the line. The architect blames the contractor, the contractor blames the foreman, the foreman blames the men and they have to carry the buck. And we had a major mistake on our first job in that the precast units that we had bought, when they came on the site wouldn't fit. And for two to three days the chaps struggled to try and get these to fit, and I noticed there was a big Y painted in black on the back of the units, and I thought, I wonder what that means, so I went back to the office and rang the firm and spoke to the managing director and he said, 'Oh, my God, that means they are designed for York University.' And I gulped and said, 'Well this is a little primary school in Nottinghamshire, not York University.' And I said, 'This is going to cause major problems.' He said, 'It doesn't matter. I'll send the new ones out and when you have taken those down, I'll collect them again.' I said, 'Not so quick, that won't do. My men work to very high standards and they have been told they don't have to carry the can for other people's mistakes.' And he said 'So?' And I said, 'So you will come on site, and I will call a site meeting and you will apologise to my workmen for the mistake you have made.' He said, 'I'm not used to doing that.' And I said, 'No you're not, but you're jolly well going to do it because I'm not paying you until you do.' So he said OK. And he was a nice guy. He came in his huge BMW and all the rest of it and drove onto our muddy site. I said to the chaps, 'Come on over, this is Mr X, he has got something to say to you.' And they all gathered round looking rather sheepish, wondering what this was and I said, 'Mr X is the manager of the company who have made all these concrete units with Y on the back. He wants to speak to you.' And he said, 'I am very sorry, we have made a mistake. It is my fault entirely and I will make sure that none of the costs fall with you and I know you are going to have to take them down and it is hard work and it is cold and the middle of winter and all I can say is I am extremely sorry I made a mistake.' And he went off the site and they said, 'I've been in this industry for so and so years and I have never heard a boss apologise before and say I'm sorry . . . .'

Skill and the CLASP system

The RSM project provides evidence of the practical and ideological contradictions of introducing industrialised methods of building to the construction industry. The use of CLASP components entailed an assembly rather than traditional building process to erect the main structural frame, cladding, windows, partitions and roof. The services, however, remained traditional in their form and installation, the finishing trades of painters, decorators and tilers were still involved and the foundation slab consisted of poured *in-situ* concrete requiring formwork made on site. While Henry Swain celebrated the introduction of factory-produced components and their 'operational assembly' on site, he did not downgrade the skills of the men putting the building together, instead he praised the abilities of the 'craftsmen joiners' who were able because of their training to turn their hand to most of the jobs on site.[18] On an RSM site, the joiners fixed and assembled the steel frames, roof decks, concrete cladding and steel partitions as well as the usual woodworking jobs. Demarcation disputes did not occur: perhaps understandably as the project was a small offshoot of the Architect's Department, and seen as an experimental unit.[19]

The project encouraged in the building workers a general knowledge of construction, teamwork, ability to read drawings and plan their own work, accurate setting out and the ability to respond to and overcome problems. This was radical in comparison to what most construction employers wanted from their employees. Industrialisation, for many, was an answer to the skills shortage precisely because it supposedly did not require any skill from the workers. Swain, however, understood the change in site processes as necessitating a wider and different range of skills that still needed the breadth of training traditionally found in a good apprenticeship scheme. In the account written for the *Architects' Journal* he included a photograph of a joiner's work station showing the wide range of handtools needed to fit a small baluster of steel, with a timber handrail, into concrete steps. An example of the complexity of the problems found on site and a very long way from the smooth clipping together of manufactured, universal components dreamed of by the members of the Modular Society in the 1950s.

The different attributes that skilled men brought to the building process were often hidden from the gaze of the architect although the men themselves were aware of the range of individual contributions made. Speed was not always the main asset that a skilled worker brought to the site and when one very slow bricklayer was threatened with the sack, the foreman challenged the architects and defended him on the grounds that he was very accurate and was always given the task of laying the foundations and putting in the footings because his setting out was so precise. His work then enabled the subsequent bricklaying to proceed at a faster rate.[20]

Alan Meikle's reminiscences point out the difference between the RSM method of building, a collective effort, compared with typical practice by construction contractors:

> [L]aying *in-situ* concrete, [was a] dirty, heavy, mucky job, nobody liked doing it and also it was a key element in the critical part in the early stages of the job, and we didn't just want to hire gangs or labourers to shovel a lot of concrete around. So we investigated the practicalities of delivering vast quantities of concrete and instead of spending six weeks laying a base of concrete in the traditional way, we came to a play whereby we could lay the whole of the concrete floor of a school in two days. We would do half one day and a week later we would do the other half. But to do that meant that everybody had to be laying concrete. So on the appointed day, everybody, carpenters, bricklayers, quantity surveyors, the lot, would lay concrete and they would start at 6.00 a.m. and a fleet of lorries would come and the pump would work and they would go through until they were absolutely exhausted . . . Everybody

accepted that because they could see that we saved a lot of money and on that particular day everybody worked on the same task and there was never any problem . . . And I think wherever there was a logic in the argument, people were willing to forget their trades. On the other hand, dealing with the professional people, I did have a serious problem with the quantity surveyors and one job I got through 16 quantity surveyors but most of them wanted to drop being professional people and get their gum boots on and start laying concrete. And they didn't do it very well and we had some huts erected by quantity surveyors and they were a disgrace. I think they wanted to play at being builders because they had such a boring job . . . So there was a terrific interaction . . .

I think there are a lot of architects – I would put it as far as 10 per cent, probably have got a very strong practical orientation and they feel frustrated.[21]

Although the design of the work sheets enabled feedback on site operations to reach the office, it was not until quite late into the project, in May 1970, that Meikle introduced Joint Consultation Meetings on a regular basis. Prior to that, meetings had been 'one-sided' exhortations on the part of senior staff members. In a sense the project management style was one of paternalistic liberalism and worker control was not really on the agenda. In the same way that the fundamental inequalities, based on class and economic power, were not addressed, neither was gender inequality. The project attracted brief attention in the media for employing female Swedish architectural students, but this was a carefully orchestrated piece of public relations.

Working on the project appeared to have a major effect on the career direction of the majority of job architects who had been, in a sense, self-selecting. Nick Whitehouse remembers applying for a post straight from Sheffield University because he had read about RSM in the architectural press. He described himself as having a practical approach to architecture, not being particularly interested in 'high aesthetics', and had always thought that a young architect had a lot to learn from building craftsmen. Nick Whitehouse, now Managing Director of Terrapin Ltd., ran the eighth and most successful scheme in Henry Swain's account, considers working on RSM a formative influence on his career. He continues to take 'advice from the guys who do the job' as 'you disregard comments from the workforce at your peril'.[22]

Although the RSM experiment generated considerable interest in the building press, it did not change accepted practice in the industry. It was a very small organisation and the buildings it erected were straightforward and relatively simple so that it could not be seen as paradigmatic for the entire construction industry and for large complex projects. However, its findings presented a considerable threat to the status quo of one the construction professions. Eradicating the Bill of Quantities and throwing out the SMM resulted in a reduction in the number of detailed architect's drawings intended for the use of the quantity surveyor. Although this generated considerable cost savings, it undermined the professional role of the quantity surveyor and in the RSM project they functioned as accountants. Placing the architect on site, acting as site manager, not only dispensed with the Clerk of Works but also, due to the architect taking responsibility for keeping the project to costs, the contractor's contract manager.

For the operative workforce, the project appears to have provided a rewarding, happy, respectful and stimulating workplace, but still without any real worker participation in the control of the building process. The tangible benefits belonged to the architects who believed in Stirrat Johnson-Marshall's advice 'put nobody between the architect and the men'. What it demonstrates very clearly are the financial and organisational benefits of decreasing the hierarchy of site management. In Alan Meikle's words:

> What we do want to avoid is duplication of management. We don't want project managers, we don't want site agents, we don't want layer upon layer of men telling people what to do. What we want is to preserve this direct communication between designer and maker.

Establishment modularity: the War Office under Donald Gibson

Donald Gibson, whose career epitomised the public service architect of the post-war era, joined the War Office in 1958 as Chief Architect when it became a civilian department and was instrumental in the move to state support of industrialised building.

Roger Walters remembered the occasion:

> In 1959 I was asked by Donald Gibson to go and join him in the War Office where he was setting up a new organisation based on principles worked out in the Ministry of Education where administrators and professionals were given equal status and shared the responsibility at each level. I found it a very exciting time. [Later] in 1963 we were suddenly taken over by the MPBW in the shape of Geoffrey Rippon – that's a story in itself which is not going on the record – and Donald Gibson was asked to become the new Director of Research and Development. He said, 'I don't know what I'm in for but will you come with me?' and so I said 'Yes', and we went and set up this new organisation in the Ministry. And we were the Ministry's sponsorship arm – the Ministry always had a sponsorship role with the construction industry. Rippon wanted his Ministry to be effective and, I'm not going into all the things we did, but we promoted industrialised building and dimensional co-ordination.[23]

When this department merged with the research and development wing of the Ministry of Public Building and Works in 1962, he became Director

General of a large and influential organisation. A keen promoter of dimensional co-ordination as an aid to greater efficiency in the building industry, he looked forward to the day 'when building can equate with other sectors of our economy as aircraft and engineering'.[24] While still at the War Office, Gibson had introduced the use of CLASP on new building contracts and pioneered research on the NENK building system. From 1963 onwards, a series of design guides on dimensional co-ordination for industrialised house building had been published by the MHLG.[25] These all emanated from Donald Gibson's research and development department and were part of Geoffrey Rippon's drive to industrialise the construction industry. However Gibson brought with him some of the ethos of the school building programme as evident in his approach to the development of the NENK method. This was a 'light and dry' system based on space frames and built up of prefabricated steel pyramids promoted as, unlike many industrialised systems, giving full responsibility for aesthetics to the architect. A report of a packed meeting at the RIBA in 1964 where the newly built barrack blocks at Maidstone were presented claimed that:

> Perhaps the most encouraging contribution to the discussion came from Mr Price of the NFBTO who warmly praised the industrial relations at Maidstone: his district officials had been consulted at every stage including seeing an initial mock-up at the War Office. If anything there had been even greater craft skills needed on this job than on traditional work.[26]

It was under the aegis of Donald Gibson at the War Office that Farmer and Dark gained the contract for the first fully modular buildings to be built on an industrial scale: the Little Aden Cantonment (1960–63). The Farmer and Dark practice, established in 1934, employed a range of in-house construction expertise with an emphasis on engineering as an equal partnership, and had produced a number of significant buildings in the preceding decade.[27] One of these, Marchwood Power Station, was a much-lauded post-war building, described by Nikolaus Pevsner as one of the 'best pieces of industrial post-war architecture in Britain'[28] and included in Trevor Dannatt's *Modern Architecture in Britain*.[29] The American critic, G. Kidder Smith found it, 'a precise statement of envelope, the envelope closely reflecting the contents, set off by a lean and virile play of ancillary units . . . This power station along the shipping lanes to Southampton is among the most impressive in Europe.'[30] In particular the vertical saw-tooth glazing and aluminium sheeting of the turbine hall were praised for both allowing in daylight and, by night, illuminating the hall via the internal lighting system. Andrew Derbyshire was one of the young architects responsible for designing the glazing system:

> Like Bruce Martin said, we need to build, we need to harness all the resources of science and technology if we're to produce an architecture

that's fit for the welfare state. We were a bunch of students at the AA who actually took this very seriously, and we believed in the modulor, which was an opportunity to use a dimensional co-ordination which would also express, through the Fibonacci series, express the golden section, so it was a wonderful synthesis of aesthetic perfection and technical expertise, through standardisation, it was the gift from God! Actually, we never managed to make it work [laughing]! Very difficult to handle, those Fibonacci series . . . But we wrote a number of articles in *Plan* about the merits of prefabrication, and we did an article on the Hertfordshire schools for instance, and that sort of thing, and I emerged imbued with a sort of crusading spirit to prefabricate, to bring the construction industry into the twentieth century. I don't know how we persuaded Frank, or the civil engineers for that matter, to endorse light cladding, aluminium and glass, and so on, maybe brick shortage was instrumental. Anyway, we managed to persuade them and they gave us free rein to do a light cladding system for the station. We designed a system of three-dimensional cladding struts, triangular in section, which would be glazed on the top surface and had aluminium sheeting on the return surface. So it was a sort of section, which provided lighting, natural lighting, for the plant, the machinery spaces, and was very quick to put up, and very cheap. I think Marchwood was probably the first of the prefabricated cladding power stations.

Modular Society members visited Marchwood Power Station, Southampton, in 1955 and Bowater Paper Mill, Northfleet, in 1956, neither of which had been designed to the 4-inch module.[31] Marchwood was planned on a 13-feet grid with a vertical module of 9¾ inches derived from repeated halving of the grid.

However, it was the Little Aden Cantonment that particularly excited Modular Society members. Alan Diprose wrote in 1962 that it was an indication of how quickly modular ideas had progressed in that 'a new concept in building [was] to be realized in the form of a complete town within ten years of the first statement of theory'.[32] It was introduced to *Modular Quarterly* readers as 'the largest fully modular project so far undertaken anywhere in the world. By "fully modular" is meant not only a modular planning grid but modular components throughout.'[33] The other key factor was that it was designed throughout on the basis of the 4-inch module, which at this point had still not been agreed by the BRS or the Ministry of Works, but in this particular project was being endorsed by the War Office. The project, for which Bruce Martin acted as a consultant on modular co-ordination, comprised over 1,000 buildings. These included housing for officers and soldiers, a school, petrol station, shop, recreational facilities and military training facilities designed to fit into a two and a half mile strip of desert located between a lagoon and the foothills of the Jebel mountain range (Figure 8.1). The two basic building types were those

The role of architects on site 149

PLAN AND AERIAL PHOTOGRAPHS OF THE MODEL.

Figure 8.1 Little Aden Cantonment, Farmer and Dark, 1961. Photographs of site model.
Source: *Modular Quarterly*.

150 *The role of architects on site*

providing small rooms for the offices and accommodation, and the large uninterrupted floor areas required by the workshops. The first type was designed to be constructed of blocks based on multiples of 8 inches (2M) with their positions described on a grid of 16 inches (4M). Centre-line grid planning was not used and components had their faces to the grid line as suggested in the Modular Society's '3 Rules for Modular Assembly'.[34] Vertical heights between floors were determined by multiples of an 8 inches (2M) module. For the buildings requiring a larger roof span, a planning grid of 10 feet 8 inches (32M) using a structural system of concrete columns was positioned outside the main grid. The actual manufactured sizes of the nominal modular components were made ½ inch less all round than their modular size. For example, a timber framed window of nominal size 4 feet (12M) × 2 feet 8 inches (8M) was made at 3 feet 11½ inches × 2 feet 7½ inches. The design of the windows themselves was highly rationalised with only one bead type for the entire range of windows with a range of standardised components used throughout the scheme. Again, as in the case of Marchwood, the scale of the enterprise enabled an internal system of standardisation to operate with the blocks being made on site and many of the components imported.

The working drawings for components were also standardised. First, a sheet, which covered the whole range of types and nominal sizes available for a particular component, was drawn for the use of the architect. Second, ranges of components of a similar type were drawn to half-inch scale and actual sizes. Third, a sheet divided into two parts: on the left-hand side, full-size sections of a component and, on the right-hand side, a sequence of drawings showing the assembly and fitting sequence of the component was produced (Figures 8.2, 8.3). These last drawings were devised for site work, the full-size drawings did not show dimensions but related the component to the modular grid, enabling the site worker to check components against the drawings for size.

One of the site architects on the Little Aden project was Alan Crocker. He was introduced to the use of modular design at Farmer and Dark by John Weate, and had high regard for the job architect, Bill Henderson, who had been one of his tutors at the Regent Street Polytechnic immediately after the war. By 1955, Farmer and Dark included two senior architects who had previously worked under C.H. Aslin on the Hertfordshire schools programme, Bill Henderson and J.T. Pinion, as well as John Barton who had assisted Eric Lyons on the Span housing schemes. In the late 1950s, these three together developed the A75 prefabricated timber system for school building.[35]

Crocker regards the drawings as one of the keys to the success of the scheme:

> The secret of the success of the project was a carefully thought out drawing and specification system which really meant that local labour did not

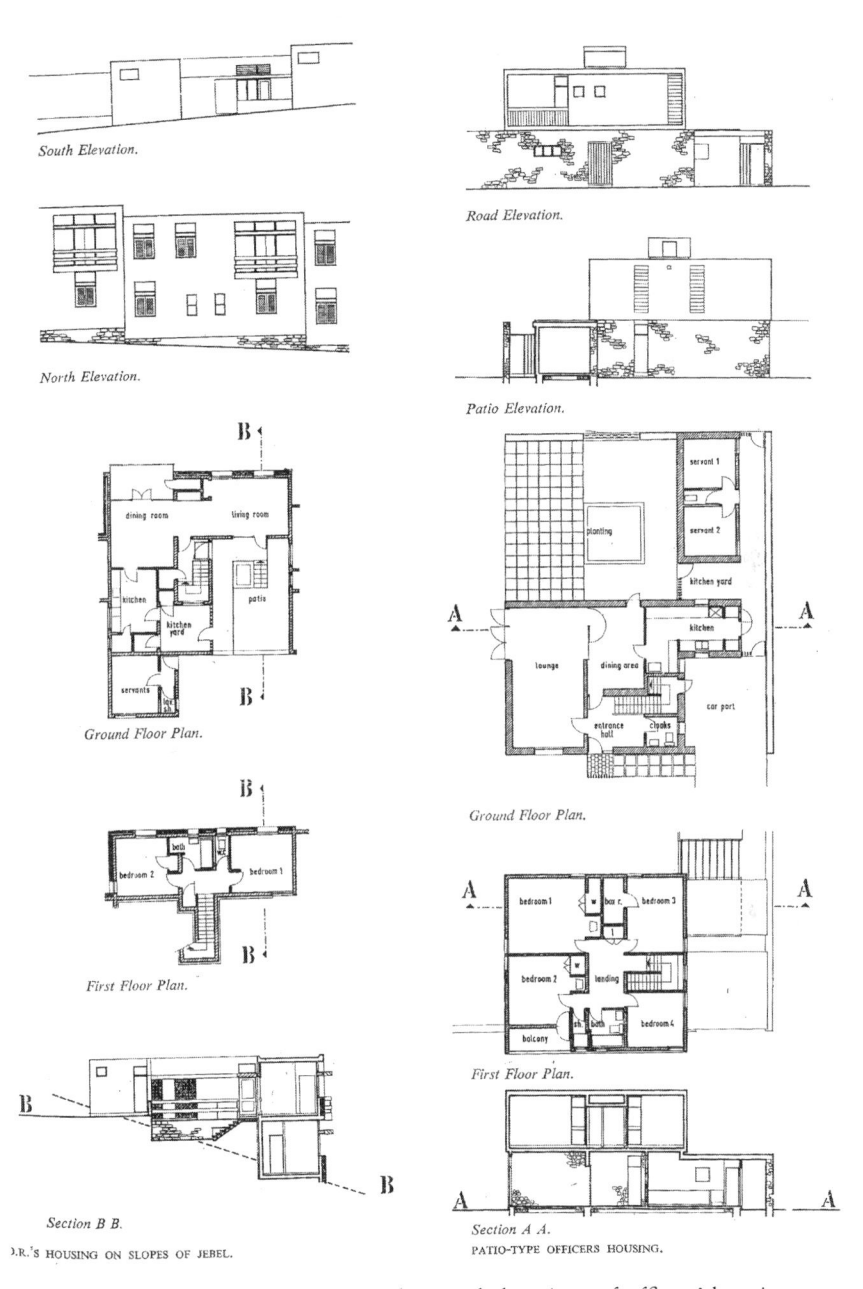

Figure 8.2 Little Aden Cantonment, plans and elevations of officers' housing
Source: *Modular Quarterly*.

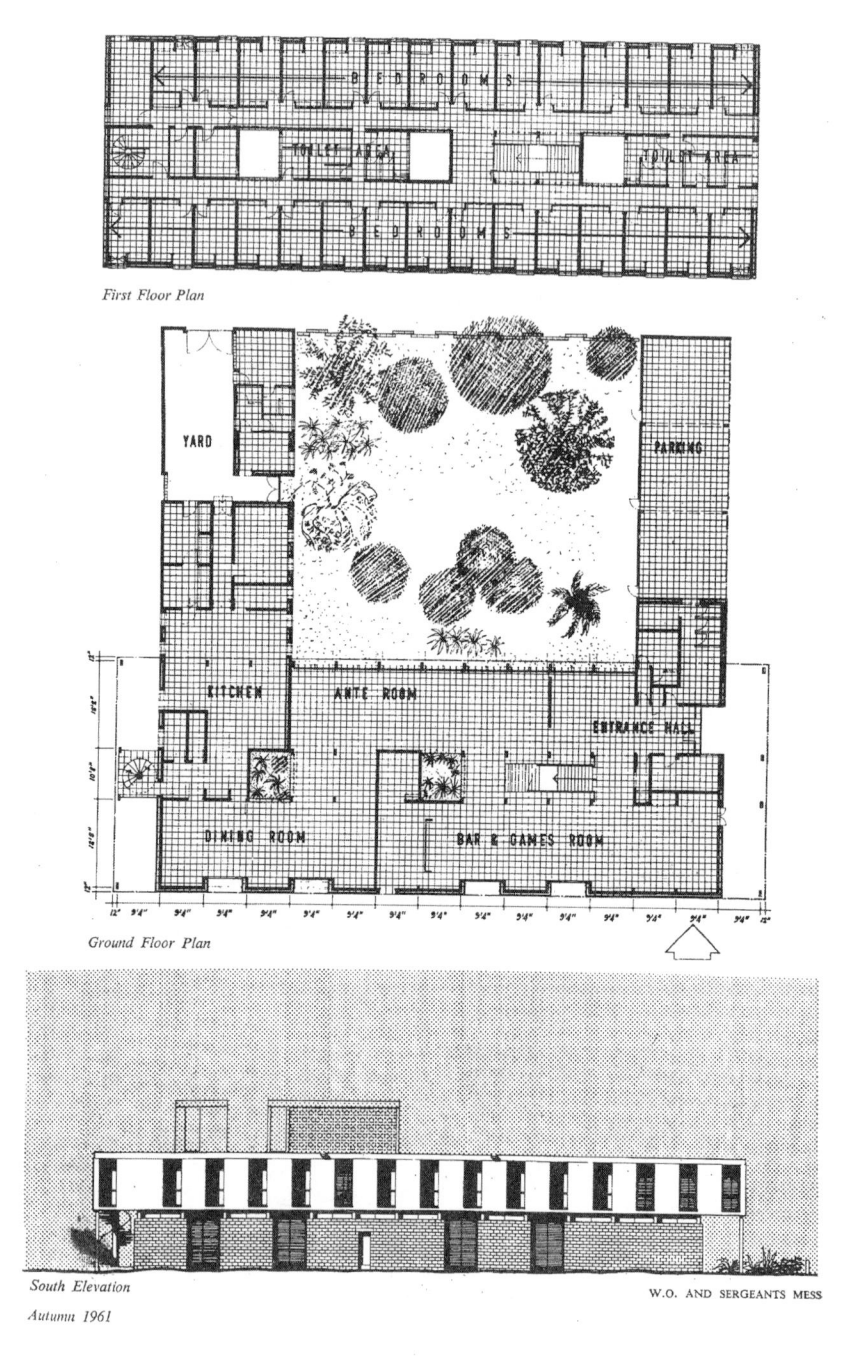

Figure 8.3 Little Aden Cantonment, plans and elevations of sergeants' mess
Source: *Modular Quarterly*.

need to refer to dimensions or deal with complicated specifications, they just counted squares on gridded paper and selected components from books of very clear drawings. It was an economical design process partly because of the sheer scale of the project using, on nearly all buildings, quite a modest range of components. Because of the use of standard components, some made on site, some in the UK, there was no need for operatives to wrestle with complicated specifications. For example, the joinery manufacturer worked to a specification for construction and finishing of components. The site operatives were, as on most sites, skilled, semi-skilled and labourers. Even the semi-skilled could work accurately and comparatively quickly because of the careful design of components and the splitting of drawings into location, component and assembly drawings . . . Site mechanisation included craneage, concrete mixers, etc. Block making was quite sophisticated, and there was a fair bit of mechanical transport for moving materials.[36]

The other factors, in Crocker's opinion, that led to the success of the scheme were extensive and tightly scheduled pre-contract design and 'an excellent team of designers' at Farmer and Dark. The production of drawings, clear enough to be used as tools of communication on site without the usual re-interpretation by a foreman, created a huge advantage to site efficiency. They also demonstrate an approach to industrialised building that recognised the role of the operative in site assembly by giving accurate information on precisely how and where the building component was to be located and assembled, and acted as a tool for direct communication between designer and operative builder. Crocker went on to write a handbook on modular design, which, unusually for this type of manual, included advice to architects to listen and 'learn much of value from so-called labourers' and reflected that:

> No matter how sophisticated the methods of construction, how well prepared the drawings, specification and schedules, how much work is prefabricated in factory conditions, it is the men on site who, for many years to come – until, in fact the advent of a new technological, industrial revolution – will be the men without whom the rest may as well not put pencil to paper.[37]

Both these accounts suffer from not having the voice of the operatives themselves recounting their side of the experience.

9 The nature of work in the construction industry

> There were no toilet facilities provided and we had to make our own arrangements. One bucket of cold water was for washing our hands and talking of hands, there were no protective gloves available. Hands become extremely cold and calloused in the wintertime.
>
> Oliver Lalor, Irish labourer in the 1960s[1]

The idea that the 'shoulder-to-shoulder' experience of common hardship endured by the British population during the Second World War acted as a social leveller of society, resulting in a blurring of class divisions, is now largely discredited. Social divisions continued throughout the war, the affluent rural middle classes experiencing a much easier war than the urban working classes.[2] For those on the Left, the Second World War was the harbinger of a social revolution, with the Beveridge Report and the setting up of the Welfare State the first steps towards socialism. But socialism, at least the sort that eradicated class inequality, did not arrive with the Labour government of 1945 and Britain remained a deeply divided society.

After the Labour Party lost the 1951 general election, there followed 13 years of unbroken Conservative rule. While popular debate linked working-class affluence with a change in voting tendencies to explain the defeat of the Labour Party, sociological research into the working lives and attitudes of manual workers attempted a scientific assessment of the effect of affluence on class structure.[3]

In an overview of the research conducted in the 1960s on social class, Goldthorpe and Lockwood report how their work refuted the thesis of 'embourgeoisement'. This widely held belief predicted that, with increasing affluence, the working class would disappear into a larger middle class. Instead, they concluded, that despite higher working-class incomes, more material possessions and changes in the location and type of housing in which they lived, the working class remained distinct from the middle class. The clearest line of social demarcation in British society remained that between manual and non-manual work.[4]

Manual workers in construction

New insights into the male manual working class revealed clear hierarchical social distinctions between the 'rough' end, composed of labourers and unskilled workers, and the 'respectable', skilled tradesmen and shopkeepers. As described by Robert Roberts, 'the real social divide existed between those who, in earning daily bread, dirtied hands and faces and those who did not'.[5] Manual workers in construction were unable to hide their class position even if they had wanted to. Poor working conditions and lack of site facilities meant that their grubby work clothes publicly identified them as manual workers, unlike skilled factory workers who were able to change out of their overalls in factory changing and locker rooms before they emerged into the street. The lack of washing facilities on building sites meant that building workers were instantly recognisable in the street by their dirty clothes and were very conscious of the effect this had on other people, as this bricklayer recounts in 1949:

> We all bring a brush to work for our boots and clothes, it makes you feel bad to sit on a bus in dirty clothes and for the person next to you to give you a dirty look and then move to another seat.[6]

It was not just the workers who were disgusted by poor site conditions, the architectural press also decried the chaos and lack of facilities found on most sites. Peter Dunican, writing in the *Architects' Journal* in 1954 contrasts British practice with sites in Denmark and Sweden, building the Larsen-Nielsen system. Here site welfare was provided in electrically heated site offices and huts, described as having a standard of finish 'as good as, if not better than, the average post-war [British] house'.[7]

The author of a series of articles in *The Operative Builder* on site welfare needs (that is, the provision of canteens, shelter, water, washing, drying of clothes and sanitary arrangements) written in 1948 urged building workers to demand better conditions:

> The greatest hurdle we have to clear is of our own making, namely our reluctance to change our age-long indifference to our welfare needs ... When the majority of us become aware of the dignity of our status as building trade workers, and relate that dignity to the good things of life, then the difficulties that look so formidable today in relation to canteen and other aspects of welfare will soon be overcome.[8]

A 1951 booklet aimed at employers stated that on sites where there were more than 100 men, a fully equipped canteen capable of serving hot meals should be instated from the outset of the contract (Ministry of Labour and National Service, 1951: 19). It also contained explicit instructions on how to erect suitable drying rooms, sanitary arrangements and first aid rooms, but these recommendations were largely disregarded by employers and there was no legislation to enforce them. In the mid-1960s, on a major landmark site like the Barbican, there were no toilets, changing rooms or any welfare facilities provided in the early stages, leading to a number of walk-outs and disruption by workers until they were installed (Clarke et al., 2012).

Status

Building workers' perceptions of their societal status and the way in which their employers treated them did not change over the post-war period studied here. If anything, they became entrenched. The following account is by a bricklayer writing in 1969:

> Perhaps disillusionment comes because of the way my employers and society treat me as a bricklayer. The employers treat me as a means to an end, seldom as a person. An expendable item to be discarded like an old boot when no longer required . . .
>
> Like most bricklayers I am a fiercely independent character. I owe no loyalty to any firm or any man. I will quit a job if slighted by a supervisor or if a 'bigger penny' is offered elsewhere. I do my job well with an ease gained from years of experience. On the whole, I resent being where I am and the social position accorded to bricklayers. I might add that there is some justification for this: building workers are a rough and ready lot. Most of them are aware of their lowly social status and comments like 'You can't get lower than building worker' and 'We are scum' are often heard.[9]

Unlike the manufacturing industry, construction retained a largely casual workforce, especially among labourers. Many workers were employed for the duration of a contract and then laid off, continuity in holiday and sick pay being achieved through a stamp scheme introduced after the war. Employment conditions in construction were poor compared with the higher-paid, permanently employed car workers. A survey of building operatives undertaken in 1967 found that 79 per cent thought that permanent employment with a firm would be a 'good thing', but exactly the same proportion also thought that freedom to move around was a 'good thing'.[10] The researcher concluded that the idea of freedom was cherished rather than the reality. The limitations of the questionnaire, however, conflated 'permanent employment' with improvement in employment conditions, that is, provision for a pension scheme, sick pay, and holiday pay, so it is understandable that such a high proportion of operatives approved of permanent employment. The idea of 'freedom', however, was likely to be interpreted as a sense of autonomy, a central component in the self-image of both the skilled craftsman and the labourer but which does not preclude a collective class identity.

Freedom and individualism were the defining attributes of the gangs of 'travelling men' engaged on large civil engineering schemes in the post-war years. These construction labourers were fiercely independent in their working lives, frequently moving from employer to employer and considering it a loss of dignity to stay too long with any one contractor. To be called a 'Wimpey's Man' or a 'Costain's Man' was an insult and navvies were proud of having no allegiance to any firm, or even a trade union.[11] It was these characteristics that made the study of construction workers and labourers in particular so problematic for sociologists in the 1950s and 1960s when the dominant paradigms in sociological theory were addressing the structure and collective nature of groups.[12] However, in 1953, a detailed 'participant-observer study' of navvies working on the construction of a hydro-electric dam in Scotland was made by J.M. Sykes although he did not publish his findings until 1969.[13]

Sykes recorded the mutual contempt that workers and management held for each other. To the navvies all construction industry employers were ruthless in their attempts to exploit them and had not the slightest interest in the welfare of their workforce. The management and staff interviewed on the dam and other construction sites in the study made no attempt to hide their contempt and described the men as 'animals' and 'people outside society'. The men asserted their freedom and independence through constantly changing employers, and, if a site did not appeal to them, by 'jacking', that is, handing in their notice suddenly and walking off the site at short notice. The navvies praised those among them who worked hard while those who dodged work were termed 'latchicoes' or layabouts. Hard workers were seen to deserve their individual higher earnings gained through bonus payments.

The navvies' high valuing of individual freedom and independence can be interpreted as a response to the casual employment conditions and competition for jobs endemic in the civil engineering industry. Their lack of interest in trade unions was notorious: these were regarded not as their own organisations but as external bodies with the potential to interfere with their independence. Sykes concluded that navvies were different from the workers of other industries in their thorough separation of their home lives and work lives. They were acutely aware of the low status their occupation was held in by society, and dissociated themselves from this by regarding the work as 'temporary'. Some of the Irish maintained their dignity by leaving the industry at intervals to return to agricultural work, networks of familial relationships at home and a very different status to that experienced in Britain.

The working lives of navvies or as they were officially designated, general construction operatives, can be seen to exemplify, in an extreme form, many of the characteristics of construction work enjoyed by skilled workers. In 1946, a survey of 400 construction workers found that the most frequently cited reason for enjoying their jobs was 'the sense of freedom and the open air life'.[14] These findings were replicated in the Government Social Survey of 1967,[15] and also in 500 essays collected from new apprentices by the

BRS.[16] In all these accounts, alongside 'freedom' was the desire to escape 'factory-type restrictions' in their working lives by the variety of work that construction offered. It was precisely the opposite of this that building employers were keen to implement with the introduction of industrialised building which involved higher levels of supervision and assembly line methods originating in the factories of manufacturing industries.

In Britain, alongside Goldthorpe's pioneering work on car assembly workers, studies were made on male manual workers in the older industries, including agriculture and mining.[17] Most of this work considered the effects of industrialisation and modernisation on the workplace and the social backgrounds of the workers. Despite its size, capacity and economic importance, the construction industry was largely ignored in these post-war studies and no major investigative study was attempted. Although the construction industry produced products that visually represented, in the form of tower blocks and New Towns, the idea of mass production equally well as the identical cars produced on the shop floors of Coventry and Cowley, the similarities between manufacturing and construction were superficial. Whereas the output of a factory-based industry like the car industry could be directly connected to its manufacturing processes, the production processes of the construction industry with its immensely diverse range of outputs were far more complex to interpret. The construction workforce, too, did not 'fit' well with current sociological interpretations and perhaps contributed to its marginal interest to sociologists. However, working life is described in literature and oral histories with far more vibrancy than found in sociological accounts.

The experience of work

There is no doubt that the working conditions on most post-war building sites were appalling. Ferdinand Zweig found that ex-servicemen were used to higher standards of welfare in the Army compared to the working conditions found in the building industry in the 1950s.[18] Protective clothing did not exist and building workers were responsible for providing their own work clothes. Brian Behan, brother of the playwright Brendan, in his autobiography, describes working as a labourer in 1950s London, pile-driving in the middle of December on a site 'like a swamp' where the workers 'staggered and floundered about like ducks on a muddy bank'.[19]

> What prodded me was the appalling mud that caked your hands and splashed all over you. Naturally, there was nowhere to wash. The best you could do at the end of the day was to get a bucket of hot water out of the compressors and wash in that. We started at 7.30 in the morning, and went on till 7.30 at night. We finished as we started by the light of flares. Then it was cold. My God, how cold it was. I smoked fag after fag, more to keep myself warm than anything else.[20]

It was frequently labourers who bore the brunt of brutal management techniques used by the general foremen in charge of hiring and firing at site level. The post-war government policy of full employment never quite reached the labourers of the construction industry and unemployment remained at around 10 per cent over the three decades studied here. In 1948, a change in the working rules governing employment in construction ensured that, instead of two hours notice at any time of the week, termination of employment was only allowed on Fridays. This was used as a management tool in an industry where employers did not disguise their contempt for the workforce and where the prevailing ethos of management consisted of instilling the fear of unemployment. The following recollection is from Don Baldrey, a 68-year-old former crane driver who went on to run a successful plant hire firm and spent the whole of his working life in the industry:

> The construction industry was barbaric, my first employment with Higgs and Hill was as a plant operator . . . probably about 45 years ago and the site was down from Belmarsh Prison, on the marshes down between Woolwich Arsenal, Abbey Wood way. And there was a big labour force, hand digging trenches and that kind of thing, and on a Friday afternoon at 2.00 p.m. the general foreman would walk round in his brown boots and his cord trousers and tweed jacket and his flat cap and he would say. 'You, you, you . . . ' ten men always ten, 'Right, get your cards.' And they would say, 'What have I done, governor?' and he would say, 'Never mind what you have done, get your cards.' Ten men every Friday and a whole new bunch in on the Monday. He wouldn't fire any of the key guys, the boiler men . . . these are labourers. We had a workforce of getting on for 100 so he would fire 10 per cent each Friday. 'I'm the bloody general foreman, mate, don't come any dog with me or you are off the bloody site, don't even look at me.'
>
> It is one way of managing men, I have never thought much to it. I've managed my men rather different.[21]

Although the general foreman was in charge of overall hiring and firing on site, day-to-day management of the labourers working in gangs of three to four men was the responsibility of a ganger. Some still dressed in the traditional clothes of a navvy, and were renowned for the brutal manner with which they interacted with the men in their charge. The wages system that gave bonuses based on the work of individual gangs encouraged bullying. 'Many a novice complainer in an Irish work gang was shut up with a belt across the mouth from an Irish foreman intent only on bonus profit for himself.'[22] Brian Behan describes the ganger he was working under on the Festival of Britain South Bank site in 1950, and even allowing for the artistic licence in Behan's autobiographical writing, it is evident that there was a large gulf between the supervisors and the supervised:

> Our ganger was a vicious, ignorant pig. He dressed himself up like old Moleskin Joe. He wore a hat, knotted scarf, and hard navvy cord trousers. He even chewed tobacco. A swine of the first water. Sooner than leave us alone and unwatched, he stood on the top of the trench and pissed where he stood. His signal for the end of tea break was to pick up a brick and toss it onto the roof of the hut.[23]

'Moleskin Joe' is a reference to Patrick MacGill's popular novel about an Irish navvy's life on the road early in the twentieth century.[24] Although this way of dressing gradually disappeared in favour of the ubiquitous donkey jacket and rubber boots, it was still prevalent in the 1950s and into the early 1960s when Donall MacAmhlaigh describes the foreman ganger he encountered as typical of the type:

> Interested in nothing but the work itself and the pub, they dress in a manner they accept as being laid down for their kind – moleskin or corduroy trousers (or knee breeches tied at the knee with a cord), a scarf tied round their necks and a hat or a cap that is only taken off going to bed, if at all. Most of them are Irish but some come from the North of England also.[25]

This 'fiercely independent character' owing 'no loyalty to any firm or any man' reveals the same identity in relation to work as the navvies in Sykes' study. There would appear to be no difference between the skilled and semi-skilled in the way they perceived the employer's treatment of them as one of pure exploitation. It could be argued that this mode of behaviour is an unintended consequence, for the construction worker, of the employer's contempt. Where the relationship is less openly exploitative, with the workers held in higher regard and enjoying better conditions of employment, then loyalty to the 'firm' becomes equally characteristic. This was the case in the RSM experiment described earlier, for the DLOs attached to various local and metropolitan authorities and in many large paternalistic, private sector firms like Laings and Unit Construction of Liverpool.

The importance of a sense of autonomy and dignity for the male manual worker, his distance from the contractor's foreman and his disgust at the working conditions in the industry are aptly summed up in the following passage by Brian Behan:

> For me, the struggle on jobs has never had anything to do with wages. To me the fight to prevent indiscriminate sackings seemed infinitely more important. The dignity of the individual is not worth a curse if he can be thrown out at the whim of some foreman. How can you feel great if you have to gaze down a filthy manhole, ten foot deep, while you try to use it as a toilet? What sort of stinking world is it that denies the manual worker the same rights as the clerical?[26]

Gender perspectives

The virtual invisibility of women in the construction workforce contributed to the myth that the work itself was inextricably bound up with notions of masculinity. Heavy, hard dangerous work not suitable for women fitted well with the prevailing assumption, across all sections of society in the 1950s and 1960s, that women were inferior, both physically and intellectually, to men. Manual work in the construction industry was, and still is, work that is central to hegemonic masculine identity, an identity that needs to be in opposition to notions of femininity for its very existence. Given the prevailing ethos that construction work was too heavy and dirty for women, modernisation and industrialisation using light weight prefabricated components could have been used as a central argument for employing women, in large numbers, in the industry. That it was not is partly due to the power of the construction union's protectionism of the right of their male members to a 'family wage'. The neglect of their women members' rights is a consequence of the wider social policies in evidence at the time. It was not until 1955 that a Women's Agreement was incorporated into the Building Industry Joint Council machinery for wage claims at a point when women manual workers numbered less than 1 per cent of the industry. The President of the NFBTO, regretted that, 'The principle of equal pay has not so far invaded our industry, especially as its practical effort should not prove a burden in view of the small numbers involved.'[27]

Women architects were also in the minority, as were women in all the professions, during the post-war period. In 1955, there were 369 women members of the RIBA, making up 5.8 per cent of total members.[28] By 1965 although greater numerically, at 549, the percentage of women members had declined to 4.7 per cent. The body responsible for registering architects, ARCUK, did not keep a gender breakdown of its members in the 1950s but there are records for the 1960s (see Table 9.1), which show the same general decline in the numbers of women architects as found for the women building workers. The social policies of this era combined with labour market discrimination against women worked equally well across class boundaries. However, the identity of 'architect' was less closely associated with societal notions of masculinity than the identity of 'building worker'. There was nothing personally contradictory for women in considering a career in architecture even if it was a highly male-dominated profession.[29] For those women who managed to ignore or accommodate the social pressure to devote themselves to their families, professional life was rewarding. Nadine Beddington, company architect to three shoe retailers, interviewed in 1966, found that, 'in the Men's World of the Building Industry prejudice seems to dissolve directly one is personally involved'.[30] Ms Beddington found site supervision an 'absorbing and exciting' part of architectural work, and in common with the four other women interviewed in the same article, did not feel at any disadvantage in relation to the building workers she was supervising.

162 *Nature of work in the construction industry*

Table 9.1 Women architects newly admitted to ARCUK register, 1960–69

Year	Number of women members	Total number of members	Women as % of total
1960	38	671	5.7
1961	29	886	3.3
1962	35	1022	3.4
1963	17	466	3.6
1964	26	608	4.3
1965	27	745	3.6
1966	21	686	3.1
1967	32	743	4.3
1968	34	809	4.2
1969	35	795	4.4

Source: M. Fogarty, *Women in the Architectural Profession, 1978* (London: Policy Studies Institute).

Instead she argued that it was the working practices found in architects' offices that created the most obstacles for women. These should be run on less rigid lines including part-time and 'unconventional time' to allow women to manage their childcare responsibilities with work. She was pertinent in comparing this with the amount of time senior partners spent out of the office on committee work, research teams, conferences and seminars with no obvious detriment to the practice.

The proportion of women new entrants to architecture schools was 8–9 per cent in the mid-1960s increasing to 12–13 per cent by the mid-1970s, but these proportions were not reflected in the numbers qualifying as architects. There were, and still are, a large number of women who do not take their final professional qualifications.

The social composition of women architects was overwhelmingly middle-class, more so than the men. By 1960, over 70 per cent of women entrants to architecture schools came from independent or direct grant schools compared to 46 per cent of men. According to a study made in 1978, the only period when there were a significant number of women from manual and clerical backgrounds was around and just after the Second World War and up to the 1950s, but no explanation is offered for this.[31] It is possible that this cohort joined offices in wartime, without any formal qualifications, as architectural assistants or draughtswomen and then studied part-time to qualify.

Class position and work: architects

Most sociological investigation in the post-war period focused on working-class life and work with very few attempting any analysis of the middle or upper classes. There remained two popular interpretations of the British class system; first, that it consisted of three strata: upper, middle and lower, and, second, from a working-class perspective, there was 'us' and 'them', these two groups being broadly synonymous with the two major political parties.[32]

A third notion of class was also popular in the 1950s – that of 'The Establishment', however, this group was very difficult to define. Hugh Thomas, editing a book with the same title in 1959, included chapters covering the City, public schools, the Army, the civil service and the BBC. The expansion of the civil service due to the massive demands of the new Welfare State had created a much larger, higher echelon of senior civil servants – the mandarins. They personified the 'old boy network' of popular imagination and were perceived as having extraordinary power over the implementation of political decisions.

Anthony Sampson undertook a more thorough investigation into the nexus of relationships that might configure a discrete stratum of society identifiable as 'The Establishment'. In *The Anatomy of Britain* (1962), he interviewed thousands of 'top' people from industry, politics and the civil service in Britain. His analysis of recruitment to administrative grades in the civil service revealed that it was Oxbridge-dominated, and that between 1948 and 1956, 50 per cent of young recruits were Oxford graduates and 30 per cent from Cambridge. He concluded, however, that the idea of the Establishment 'is a mirage. A cluster of interlocking circles. There is no centre. The substance is made up not of groups of powerful men with vested interests but by vague inherited climates of loyalty, habit and technique.'[33] But a 'cluster of interlocking circles' is a very apt description of the social world of work for many modernist architects in the post-war years and in particular those who became senior salaried architects within the civil service.

Architects, as a profession, were firmly positioned within the category of middle class, whether defined by government statisticians or Marxist sociologists. In 1953, 49 per cent of students in architectural schools on full-time courses had private means of support.[34] Mark Abrams' survey of architects in 1964 revealed with greater precision the social class the profession was drawn from, with 71 per cent of architects having been to either grammar or public school.[35] There was still substance to the idea of the 'gentleman architect' in the post-war years. Christopher Gotch, writing an account of his working life for an anthology in 1969, after qualifying in 1953, was aware that he was subject to the collective illusion of the architect as some kind of disinterested hero.

> I had been led to believe that I was joining a profession, admired by the public, whose members were gentlemen first and foremost, then artists; men who possessed individuality and power, men who ruled their clients as they ruled site operatives, men who were arbiters of taste and who never stooped to criticise their colleagues. It was understood that the

prime object of the architect was to build to the greater glory of God and that the rewards were sufficient compensation for this privilege, besides which, gentlemen did not lower themselves to haggle over money nor involve themselves with the sordid world of commercialism.[36]

Although architects were able, when necessary, to define their difference from the more orthodox professions by describing themselves as artists as well as technical experts, they experienced a severe crisis of social status when the salaries of the other professions were published early in the post-war period and it became public knowledge that their earnings were far less than lawyers, doctors or even engineers (see Table 9.2).

Architects' aggressive defence of their position in the hierarchy of building occupations is understandable, therefore, in relation to their relatively low salaries. The RIBA was opposed to extending the education of builders into the universities and also the provision of degrees in building. But whether they perceived themselves as occupying a precarious position in respect of their general social status compared to other professionals is difficult to ascertain. There were far fewer sociological studies of the professional middle classes than the manual working class even though much post-war research was predicated on the premise of investigating and quantifying inequality in society.[37]

Graeme Salaman (1974), in a sample of 52 members of the North-West London branch of the RIBA, compared the characteristics and attributes of architects with railwaymen, groups chosen because of the lack of information on them and also because, typically for sociological research of this period, the researcher had a 'way in' to these occupations from having studied architecture for a year and also from having worked on the railways.

Table 9.2 Comparative professional earnings in 1955 (from the Pilkington Inquiry)

Profession	Total career earnings age 30–65 in £000 sterling	
Actuaries	105	
Barristers	92	
Solicitors	88	What a man might reasonably expect to earn if he worked full-time to the age of 65 based on 1955 earnings.
Graduates in industry	84	
General Medical Practitioners	79	
Accountants	71	
University teachers	63	
Surveyors	63	
Engineers	59	
Architects	54	

Source: Cmnd. 939, February 1960. The Royal Commission on Doctors' and Dentists' Remuneration (London: HMSO).

The results give a picture of architects consistently enjoying their work *only* while they were using the skills they considered defined them as architects, namely, creativity, design and problem-solving. Only 11 per cent gained any satisfaction from the administrative and managerial aspects of their work.[38] The majority of the architects interviewed were in jobs that, they considered, prevented them from utilising these creative skills, resulting in a high level of dissatisfaction with their professional role. Salaman interprets this frustration as due to 'the nature of architects' work expectations, acquired during training and from other members of the profession, their personal investment in them, and their inapplicability to the realities of employment.'[39]

Industrialised building systems, large-scale architects' offices and the incursion of other construction professionals into the work sphere of the architect were all blamed as major sources of work dissatisfaction. The weakened professional role of the architect was blamed by the respondents on the failure of the RIBA to provide a clear outline of the occupation and to establish an area of occupational monopoly. Despite the high levels of dissatisfaction with their work, typified by the following response, as being due to, 'red tape, petty ridiculous bye-laws, the hours you have to spend on administrative work, the moronic clients', the majority of the architects interviewed remained optimistic about their futures in the profession.[40]

Most of the architects in this sample thought that British architects had a slightly lower social status than the other professions unlike in other countries; this was again blamed on the comparative weakness of the RIBA. But by stressing the social importance of their work, architects laid claim to a higher occupational status. This was reinforced by the majority of architects working in the post-war period believing that architectural form and space *directly* affected human behaviour. In one of the few articles about architects published in a sociological journal, Alan Lipman strongly contested this widespread belief in 'architectural determinism'.[41] He interpreted its prevalence as symptomatic of the way architects were re-evaluating their role in relation to a rapidly changing industry and increased distance from the client. He pointed out that for an occupation rooted in its craft origins, the tensions of becoming familiar with the techniques of mass production, modular co-ordination, critical path analysis, etc., were mitigated by a belief that architects were 'managing' social relationships through their designs.

Despite the high levels of job dissatisfaction and the identity crisis in their professional role, using Bourdieu's framework of social class, architects maintained their social status, in spite of their relatively low wages, by being in a position of high cultural capital.[42] Abrams' social survey identified the leisure activities of architects as concert and theatre-going, art-gallery visiting and, for over a third of the sample, the regular purchase of a drawing or painting. Such recreational pursuits placed architects in the higher echelons of the middle class despite their lower incomes.

The majority of architects in this study believed that they had a certain intellectual 'style' – a clarity and decisiveness developed in the nature of their work and related to the consequences of making mistakes. Many of them also believed they would be able to recognise another architect in a crowd by the clothes he (there were no women in the sample) wore. The distinctive elements of an architect's dress in the late 1960s and early 1970s were: bowtie, coloured shirt, grey suit, bright socks and suede shoes or boots and reproduced accurately and wittily in Louis Hellman's cartoons of the profession.

Salaried architects

At professional level, and within the construction industry, the issue of control was paramount. The architect's position at the top of the hierarchy of the building professions appeared tenuous in the face of the advances of industrialisation. This battle was fought with particular ferocity by the ranks of salaried architects employed in the civil service and in local government, although with what they perceived as only partial support from their professional body, the RIBA.

In 1949, 43 per cent of all working architects were employed in public service,[43] but as early as 1928 the RIBA had set up a Salaried Members Committee to serve the needs of this group of members. Dissatisfaction with the effectiveness of this committee prompted a number of heads of public departments to form the Official Architects' Association in 1937. Although initially completely independent of the RIBA, this was short-lived and the Association and its members were soon persuaded back into the RIBA where their interests were served by the Salaried and Official Architects (SOA) Committee formed later in 1947. This committee was reasonably effective in maintaining the superior status of architects working in the civil service. However, there was a high-handedness in the way it pursued its aims in relation to other construction professionals. For example, in 1951, the Engineers Guild proposed a meeting with the RIBA Council ostensibly to 'promote goodwill and co-operation between the professions' but primarily to discuss the subordinate role of engineers in the Ministry of Works.[44] They argued that despite the greater volume and higher value of engineering work compared to architectural work undertaken by the Ministry, the Directorate structure placed architects in overall control of all work even though there were fewer architects employed than engineers. While acknowledging the historical reasons for this bias, in that the Ministry of Works was directly descended from the Office for Public Works, the engineers sought control over solely engineering work. The SOA Committee's response was unequivocal, after dismissing the arguments on numbers employed as down to engineers being merely consultants to architects and on expenditure as being due to the expensive plant used on engineering works. They concluded:

> Committee thought it out of the question that the Institute should be a party to any negotiations or discussions which might have as their object

the withdrawal from or weakening of the architect's rightful position as the head of a building department.[45]

They resolved that Council should not agree to the proposed meeting. The Committee was less successful, however, in fending off the attacks arising from dissatisfied members within their own Institute. Whereas the position of architects in senior posts in central government ministries and departments was secure, and had precedent, their role in local government was relatively new and architects frequently found themselves in the ignominious position of being under the Borough Surveyor or Engineer. In the years of reconstruction the responsibility for new housing was in the hands of the municipal authorities. For many architects they were attractive places to work, however, in the years during and immediately after the war very few, apart from the large municipal authorities, had separate architects' departments devoted solely to design. Some architects, especially those outside the major urban conurbations, found themselves working in the District Valuer's or Borough Engineer's Department. Letters began arriving, in increasing numbers, from members employed by local government worried at their status in the local government hierarchy and the lack of parity between their salaries and those advertised for other professionals, especially medical officers.[46]

By the early 1950s many architects in municipal government employment had lost confidence in the RIBA's ability to defend their salaries and terms of employment. An angry letter from the architects' department of Great Yarmouth[47] accused the RIBA of forcing its members to negotiate independently through other bodies with no particular interest in the profession. But this letter also articulated a more serious charge: that the RIBA was failing to fulfil one of the primary functions of its charter. They suggested that local authorities employed architects because they were cheaper than using private practice. By raising the salaries of public-sector architects, private practitioners would function at a more competitive level and the RIBA could ensure that no architect would be securing a commission at a lower fee than his fellows.

But the response only confirmed the fear of most publicly employed architects that the RIBA was interested only in the conditions of architects working in private practice, 'The truth is exactly the opposite; it is the salaried assistant in private practice who is in fact the one in most need of protection.'[48] The rumblings of discontent continued as the numbers of publicly employed architects grew to an estimated 60–70 per cent of members, with complaints becoming more organised. In 1953, Leonard Howitt chaired a special report into the conditions of salaried architects.[49] The recommendations of this report were far-reaching and helped in the transformation of the RIBA into a more transparent and democratic organisation. Members of the Council were to be elected through an open ballot of all members and an even ratio of salaried to privately employed architects was to be kept in all committees. The most radical proposal, however, was that of petitioning the Privy Council to allow the RIBA to set up a trade union for architects and architectural assistants. This arose from the finding that the majority of salaried architects wanted the RIBA to represent their interests, an impossible role under the terms of a Royal Charter.

Architects already had the opportunity to join a trade union affiliated to the TUC, the Association of Building Technicians (ABT),[50] but in 1952 only 3 per cent of RIBA members paid their dues. The apolitical nature of most architects is indicated by this low union membership, affected possibly by growing anti-Communist feeling generally and the conflict between trade union membership and the still pervasive idea of architecture as a gentlemanly profession. By 1954, the SOA Committee had excluded ABT representatives from meetings, arguing that membership levels were too small to be recognised by the RIBA. The debate on whether the RIBA should become a trade union for architects carried on in the pages of the architectural press but did not result in any changes to the organisation.[51]

Those few architects who were socialists were usually members of the ABT. However, there were also those who were dissatisfied with the rigid hierarchies within the profession itself. In 1953, Professor Ian Bowen, a construction economist, was given guest editorship of the *Architects' Journal* and the subject of 'The Structure of the Architectural Profession and its Future Prospects', to investigate over the year.[52] The final recommendations of a series of investigations noted the low salaries and lack of a body (or trade union) to represent architects' economic interests but failed to address the growing dissatisfaction with the RIBA from salaried architects. They also maintained that the most efficient and productive method of office organisation was to use hierarchical working practices. This was quite out of step with the idea of team-working, the mode of organisation preferred by many government departments, and left-wing local authority and private practices, and reflected the preoccupation of the architectural establishment with the traditional private practice model of working. Anthony Cox, part of the Architects' Co-operative Partnership, made a strong riposte to this as a comment to the final article.

> Surely one of the most significant aspects of both private and public offices today is that a great deal of initiative and responsibility necessarily lies with those whom the report describes as occupying a 'subordinate position'. But they are not so much soldiers in a hierarchical army as members of a team. The team may have its hierarchy too, but if it is to function properly it must pass the ball from one to another and back again, not merely from seniors to juniors. If, in short, it was a bit more of an efficient team and less of a blundering army – the profession might be handling more than 25% of the total value of building contracts.[53]

Architect Alan Meikle, who worked throughout the 1950s and 1960s on the school building programme, recalled that military discipline as a form of organisation in design offices persisted for at least a decade after the end of the war. While working for Henry Swain in Hertfordshire, both men used to drop back into naval attitudes and slogans. The relationship with operatives was definitely hierarchical but uncomfortable interactions between those on different levels of the hierarchy could be mediated through rituals familiar to both parties from military service.

> It was a time, we are talking about 1950, when the architect was always referred to as Mr. It was quite common to find somebody on site still wearing a bowler hat . . . This is before the days of safety helmets. You might find the clerk of works in a bowler hat and again this was the service tradition that existed on the building site.
>
> The first Clerk of Works that I had on the job in Hemel Hempstead was an ex-sergeant major and his attitude to me as a young rooky architect was the sort of relationship I'd had in the Navy between the chief petty officer and junior midshipman and although he had to call me sir – he had to show me deference because of my technical seniority – but because of his vastly greater experience of course, I had to recognise that and find a way he could give me his knowledge and skill without undermining my own authority. But, of course, one is quite used to that because . . . always before you made a decision, you would be careful to ask his opinion before you did it . . . [54]

The architect was very dependent on good communication with the Clerk of Works for the smooth running of any project, and in a sense, Meikle has described an almost deferential relationship on the side of the architect. It is also very similar to that of an NCO (in this context equivalent to Clerk of Works and trade foremen) and junior officer (architect) with its mixture of servility, surliness, avuncular support and genuine warmth.[55] The Clerk of Works, however, in his turn had a far more brutal relationship with the men, his point of contact being the foreman:

> Now between the Clerk of Works and the foreman there was a considerable gulf. I remember going on to one site where there was a really tough Clerk of Works and he dragged the foreman who looked a bit like Arthur Askey, up almost by his collar and stood him in front of me and said, 'Now this here is Mr Meikle and he is the architect, and if you do what he says, there will be no trouble on this site but, mark my words', he said, 'Step out of line and if you don't do what Mr Meikle says, you'll be in big trouble with me. Now that is the relationship we like to have on this site. Is that clearly understood?' This was almost at the first site meeting so he was establishing his authority and the foreman was jolly well going to do as he was told and presumably in the army the Clerk of Works would have been a sergeant major and the foreman would probably have been a sapper. So one fitted into that mode.[56]

Within the profession, there were, of course, generational differences in outlook; many of the younger architects who served in the Second World War were permanently influenced by their experiences of wartime service, which had a profound effect on their sense of social responsibility as professionals. However, this rarely spilled over into support for nationalisation of the construction industry, worker control or worker participation in the building process.

10 Elusive connections
Architects and building workers in mid-century Britain

Separated by the vast chasms of class, social status, working conditions, pay and employment conditions, and linked only by being part of the same sprawling industry, it might be thought spurious to try and argue that architects and operatives had very much in common in the post-war era. However, there are a number of areas where there were similarities between the two groups, if not in their working lives. Both architects and operatives, compared to others in their occupational groupings, were poorly remunerated. By the early 1960s, even though architects' average salaries had increased considerably since the findings of the Pilkington Inquiry, they had not increased relative to the other professions.[1] In parallel with this, average manual earnings in construction stayed below average manual earnings in manufacturing industries, and in 1968 were less than 50 per cent of the average manual earnings in a comparable industry, shipbuilding.[2]

In the private sector, both groups mirrored each other in terms of structure when private sector construction firms are compared with architects in private practice. In 1962, nearly 70 per cent of architects offices consisted of between 1 and 5 staff, 18 per cent with between 6 and 10, and only 13 per cent with over 11 staff.[3] Manpower in the construction industry was also dominated by very small firms with 73 per cent of firms employing 1–10 employees and only 23 per cent employing between 11 and 99 employees (the remaining percentages consisted of very large firms employing in some cases over 1,000).[4] Both groups presented themselves as conscious of their social usefulness to society especially during reconstruction. Interfaces between these two occupational groups also occurred at a political level through the short-lived wartime BINC and through the ABT delegates' regular attendance at the annual NFBTO Conference. The ARCUK Education Board also included a member of the NFBTO; however, at no point after the end of the war did there exist an industry-wide body to facilitate communication between the disparate groups.

Perhaps the most enduring point of contact was sustained through discourses on the nature of craft and the role of the master craftsman in the history and development of the architectural profession. William Lethaby (1857–1931) is the influential architectural figure whose ideas directly addressed the divide between operative and architect and who also suggested solutions to reduce this distance. The main principles of Lethaby's proposals for improving 'the whole range of activities associated in the art of building' are contained in his 1901 lecture at the RIBA entitled 'Education in Building'.[5] Over three-quarters of the lecture is devoted to an historical investigation of the origin of the architect stemming from the medieval master masons and master carpenters. Lethaby points out that very little is known about these men in England, compared to the records available in France and Germany, and that the newly published *Dictionary of National Biography* contains not one master craftsman. He then proceeds to name the masons and carpenters he has discovered in his research on Westminster Abbey during the twelfth and thirteenth centuries. The effect of these references to sculptors, plasterers, carpenters and masons is to give equal weight to the skills of the craftsman compared to the designer, especially as Lethaby describes the task of designing as 'merely contrivance, the doing of work in an ordinary way, just like cooking'.[6] This refusal to elevate and separate design from production is Ruskinian in origin. However, Lethaby's socialism, and involvement with Philip Webb and William Morris, produced a modern synthesis to the ethical problem of how to produce an aesthetically 'pure' and worthy architecture.[7] For Lethaby, this was based on a rational and scientific approach to building:

> I feel it is false to set up an opposition between science and art at all . . . It is mere obfuscation and obscurantism and everything bad, this suggested opposition between art and science. I would make the teaching of architecture wholly science.[8]

His pragmatic approach to building was founded on an immediate improvement in wages and conditions for building workers. 'The quality of workmanship rests in the long run on an economic basis: the thought and energies of the workforce are now so exhausted by the wages war that they have only heart and strength left for routine labour.' Another suggestion, modelled on the role of the medieval master mason, was the presence of the architect on site and his direct involvement with the building process to the extent that the main contractor was dispensed with and the architect himself directly employed the labour needed. This he was to try out, although not in person, in his own practice, but with limited success.[9]

A large part of his proposals involved better communication between workmen and architects and respect for the knowledge held by the men:

> We may learn much about building from mixing with and questioning the men, who still hand on amongst themselves ancient traditions . . . These rough, tired men that sometimes irritate us, the 'so-called British Workmen,' are after all the true artists in building, the representatives of the medieval architects, and it is absolutely necessary that

some relations and community of interests should be established with them once more. As it is, I never go on a building which I call my own but I want to beg their pardon for my vulgarity, pretentiousness, and ignorance. It is they and only they who sufficiently know what stones are sound and set on their right bed; what cement works properly under the trowel; whether every tile in the roof has two nails, and so on.[10]

Lethaby was not a sentimental traditionalist intent on building in the vernacular and reviving rural handicraft skills. In his architectural work he used modern materials and methods of construction alongside traditional, as in the All Saints Church at Brockhampton, where a thatched roof covered the concrete vaulted ceiling.

In his pedagogical activities he advocated joint training for building workers alongside architects in the latest scientific techniques. Lethaby believed that art would emerge when rigorous scientific method was applied to architectural design by someone with the right temperament, but not necessarily trained or qualified as an architect. These opinions did not endear him to the RIBA or individual architects at a time when the hegemonic understanding of the practice of architecture was one of aesthetic self-expression within the parameters of a closed profession, but at an ideological level he was very influential in architectural thought and criticism.

Lethaby was instrumental, through his advisory work for the London Technical Education Board, in the setting up of the Brixton School of Building, an institution whose function was:

> training junior members of the trade in principles and the relationship of these to other allied trades, training potential foremen and clerks of works, the training of architects in the principles of construction and in all matters, in which in the exercise of their profession they come directly into contact with several branches of the building trade, experimental research of an educational type in connection with builders' materials and composite structures.[11]

Lethaby believed that the trade unions should take on the functions of medieval guilds, especially with regard to apprenticeship training. His legacy remained active at Brixton, under D.A.G. Reid, who, in 1955, regretted the fact that the RIBA gave 'little encouragement to the young man employed in an architect's office and preparing for his qualifying examinations by part-time study'.[12]

On the operative side, one of the individuals who carried Lethaby's ideas well into the twentieth century was Richard Coppock, General Secretary of the NFBTO for over 40 years. Coppock had been involved with the Manchester Building Guild, formed in 1920. This organisation, which included architects, clerks of works, technicians and building operatives was controlled by its members, ploughed any profit back into the guild, maintained continuous pay for its members (an end to casual working) and aimed to revive high standards of craftsmanship. After building 100 houses for Manchester City Council, on which Coppock sat as a member, the Manchester Guild merged with the London Guild to form a National Guild. After two years the economic slump and withdrawal of government contracts caused the winding up of the Guild experiment. As R.W. Postgate commented, however, it had been important in that, 'it gave the workers of the building industry confidence and showed them that they were competent to run and control the industry, if only they could lay their hands on it'.[13] The workers did not succeed in gaining any control of the industry and the detailed proposal for nationalisation presented by the NFBTO to the post-war Labour government was not implemented.

Lethaby's legacy continued into the early post-war years with the architectural press showing a regular, if nominal, interest in operative training matters.[14] A leader in the *Architects' Journal* of 1949 expressed concern over the levels of skill apprentices were acquiring when placed with small firms without enough variety in their work:

> It may seem that the supply of craftsmen is more the concern of the building trade employer than the architect, but the industry, like peace, is indivisible. We are now training more young architects than ever before: are they to be taught to design buildings to be built by craftsmen or by semi-skilled labour?[15]

This could be interpreted as an attempt to form an alliance against technical change and the in-roads made by 'non-traditional' building and prefabrication, generally assumed in the architectural press, to require only semi-skilled labour.

However, debates remained alive and potent in public lectures held at the AA. In May 1949, Harry J. Weaver, President of the London Regional Council of Building Operatives was invited to debate 'The Building Partnership' together with an architect, Howard Robertson, a contractor, Paul Gilbert (of Gilbert-Ash Ltd.), and a client, C.W. Baker of the East End Dwellings Company. R.E. Enthoven, President of the AA took the Chair. Weaver jumped in to the debate provocatively:

> As I see it, the architect, he has surrendered his position in the building industry and in the life of the community. As I understand it, the architect is a builder and the sooner he stops talking about art and artistic conceptions ... the better for the industry. If the architects would take a positive interest in the industry as a whole, I think we should do better, because the operative looks to the architect as his ally in the business of getting better buildings and a better industry.[16]

Invoking the common origins of architects and operatives in the medieval masons, Weaver pleaded for architects to take their place alongside operatives in nationalising the building industry for the common good. He explained the importance of real communication between architects and operatives, predicated on mutual respect for the role each takes:

> I feel that the architect in adopting this sort of supreme and absolute artistic role has deserted his position as a builder, and I think that the architect should now get down to fundamentals. Many architects do cooperate, but some co-operate only to the extent of coming on the job, seeing a bricklayer and saying 'Good morning'. They then go off the job, feeling morally righteous. The facts are that most of the operatives on site would like the architect to tell them what type of building he wants. After all, if you are a labourer and you are employed on a section of a job and you dig two trenches and are then sacked because your job is done, you may never see the finished building. So far as you are concerned, you have only shifted some muck. But if you were told why you were doing that, you might be more interested in the building. There are some architects who have attempted by models and photographs to make the workman understand that although his part is a small one, it is nevertheless part of the tremendous job of creating a fine building. Unfortunately, both the builder and the architect look upon the operative as a unit of energy. I think the architect ought to take an interest in how the men are feeling on the job and whether there is a good mess room and whether they are getting meals fit for building trade workers to have, and so on.[17]

Weaver concludes that proper training, structured to enable operatives to work their way up to professional level, is urgently required and allies operatives firmly with architects, as opposed to contractors, because 'although the builder has invested his money in the industry, the architect and the operative have invested their lives'. Weaver is here referring to the idea of both architecture and the work of a skilled tradesman as being akin to vocations both occupations dedicated, after long apprenticeships, to producing works of beauty with financial gain and profit remaining firmly in the background of their endeavours.

The NFBTO journal *The Operative Builder* published an account of Weaver's contribution to the AA debate in the same issue as an article on the RIBA conference, 'The Architect in Industry'.[18] Entitled, 'The Architect and the Operative', it introduced the results of recent research by an industrial sociologist on building workers. This was one of the few pieces of research, referred to in Chapter 9, which concerned the views of building workers themselves. It revealed that building workers were concerned with design:

> 'I thoroughly dislike old-fashioned houses, plans which were drawn up fifteen years ago are being used now, you cannot have heart in this sort

of work, you just try and get by.' Another man spoke of his pleasure in building so-called Swedish houses which were a new type of construction to him: he thought they were interesting to build and aesthetically satisfying.[19]

Although most of the men preferred the complexity involved in building schools, factories, and churches, many felt that in the 'present emergency', they should be building houses. A typical remark from a joiner summed many similar quotes, 'I prefer houses. I got a great kick when I saw the first lights go on in the first house I helped to build, and I thought of people living in it, and that people would still be living in it when I am dead.'. The researcher concluded by stating two points of great importance resulting from the study; building workers' pride in workmanship and their resentment of their low social status caused by insecure employment and poor workplace welfare facilities.

Many medium-sized family firms, one of the best known being John Laing, were lauded as benevolent employers who directly employed their employees, took on apprentices and operated an internal career structure. This firm was known for the high quality and reliability of their workmanship; Basil Spence refused to work with any other contractor on Coventry Cathedral. But there was no question of worker control upsetting the management hierarchy of such firms. The rest of the industry operated on a different set of principles. A plea from a contractor for higher moral standards of leadership was made in one of the AA lectures in 1953. Here, D.E. Woodbine Parish (a well-meaning, and Eton-educated builder who later joined the executive board of the Construction Industry Training Board) made a plea for architects and contractors to work together to alleviate the poor conditions in the industry so that perhaps the 'sourness and bitterness and unhappiness' of the operatives would disappear.[20] The *Architects' Journal* of 1953 made a more practical intervention in asking for the involvement of operative representatives on a newly convened RIBA Committee, consisting of representatives from the RIBA, RICS and the NFBTE, to enquire into contracting methods and the reduction of building costs.[21]

This discourse on craft and the narrowing of the gap between architects and operatives continued throughout the twentieth century, but with each airing lost some of its potency. In the early 1900s, when the RIBA was still formalising a system of examinations by which architects would become professionally registered, a system of joint training such as started in the Brixton School of Building was less idealistic than it might seem. All classes were held in the evenings on the assumption that by day operatives would be working for their employers and architecture students would be in the office. But, as the century progressed, the widening academic distance between vocational and general education accentuated the social chasm between architects and operatives. The apprenticeship system became marginalised from mainstream education and increasingly empty of theory, and few of the technical schools promised in the 1944 Education Act materialised. Lewis Mumford recognised that Lethaby's ideas were still apposite, over 25 years after his death, when he wrote in his Introduction to the 1957 edition of *Form in Civilisation* that Lethaby would have admired, 'the factory quarters of the New Towns, the interior of the Festival Hall and the prefabricated schools in Hertfordshire'.[22] However, Lethabite ideas waned and by the 1960s had virtually disappeared as science and technology took hold in architecture schools and the building process became more mechanised and industrialised.

From the nineteenth century onwards a parallel discourse based on the superiority of technology over human labour was current. An early advocate was Andrew Ure who stated in *The Philosophy of Manufactures* (1835) that, 'The most perfect manufacture is that which dispenses entirely with manual labour.' When applied to building, this technocentric approach explains much of the ideology behind the rationalisation movement in the 1920s and 1930s and some of its more fanciful offshoots. With the acute manpower shortage at the end of the Second World War, temporary prefabricated (or 'non-traditional') housing was promoted as a simplified form of construction that required only semi-skilled labour for assembly. There were a few architects, especially those with experience of the prefabrication movement in the 1930s, who were aware of the effect of changes in the building process on skills and they were usually socialists. Bernard Cox (1945), an architect member of the Association of Building Technicians, showed an understanding of the level of skill needed for prefabrication processes:

> A more scientific way of building, working to more precise dimensions, with standards of quality constantly under examination will surely call for a higher degree of skill in every trade. Taking the long view, unskilled labour is bound to be uneconomic for the more complicated forms of building which lie ahead of us.[23]

Hugh Anthony, also writing in 1945, suggested that prefabrication of houses necessitated better training of workers, together with a standard week and the introduction of more power tools to improve working conditions.[24] By the 1960s, craft and skill issues were largely subsumed under the heading of labour costs and discussed in terms of productivity and man-hours. The RIBA convened two major conferences on industrialisation, 'Architects and Industrialisation' in 1962 and 'Industrialised Housing and the Architect' in 1967, but representatives of the NFBTO were not involved in either and the voice of the operative went unheard. The architectural profession was on the defensive in terms of protecting the viability of the architect's role in the face of systems building and package deals. The response of Owen Luder was typical:

> Architects have to take a lot of knocks as to their inefficiency and inability to carry out the job for which they are supposed to trained, but

> the building industry generally in itself is far worse. It is organised on a craftsman basis when the craftsmen have largely disappeared a long time ago. It follows therefore that the industry generally must inevitably organise itself so that the labour content is reduced to a minimum, as much work as possible produced under factory conditions and cost and quality controlled.[25]

The 1960s, then, for many architects, saw a return to the sentiments of Ure, but the discourse of closing the gap between design and production continued despite the invisibility of the operative. How did Lethaby's socialist and humane ideas on joint training and joint working for architects and operatives accommodate to industrialisation? They were certainly not defunct among the architects working in light and dry, modular design. Alan Meikle recounts how frustrating the social divisions of the building site could be:

> Now the other thing that of course happened, was there was no consultation with the actual men who were making the building, none at all, and it was not done to ask their opinion as to how something should be done. That was considered a weakness, it was certainly unheard of an architect asking a carpenter or joiner for advice on how one might tackle something and that seemed to me to be patently ridiculous – particularly when you are dealing with men who are very experienced of life and of their own trade and have worked on far more buildings than you had ever worked on. To cut yourself off from this knowledge was ridiculous, and certainly in the eighteenth century never existed. I mean, architects in the eighteenth and early part of the nineteenth century were totally reliant on trade foremen to do their detailing for them in many cases and certainly to provide a lot of knowledge about how the details were going to work out. I don't think it was really until about 1840–50, when architects decided they had better become professionals and they adopted the habits and behaviours of the middle class . . . that this gulf really began to stretch out.[26]

Architects, like Alan Meikle and Henry Swain, with progressive ideas on site management were hampered by these social mores and rigid British class system but in the late 1960s addressed them in their own working practice in the radical Research into Site Management experiment described in Chapter 8.

Harsh criticisms of relations within the industry were published in the 1962 Emmerson Report.

> There is a good deal of criticism of lack of cohesion between architect and professional colleagues and the builder. The problem does not seem to arise in the civil engineering industry where there is close personal

contact between civil engineer and the contractor. In building, there is all too often a lack of confidence between architect and builder amounting at its worst to distrust and mutual recrimination. Even at its best, relations are affected by an aloofness which cannot make for efficiency, and the building owner suffers. In no other important industry is the responsibility for design so far removed from the responsibility for production.[27]

The result was that the National Joint Consultative Council of Architects, Quantity Surveyors and Builders (NJCC) commissioned research into communication in the industry, however, without any trade union involvement. The project was given to the Tavistock Institute where the dual methodologies of operational research and sociological investigation were applied, initially, to a pilot project. Its findings were published in 1963 and, unlike the later reports, include a reference to operative workers.[28] This is in the form of a piece of unfinished research, attached as an appendix. It is not referred to again in the two further publications that result from the project, which became known as the Building Industry Communications Research Project (BICRP).[29] However, this unfinished appendix provides a rare insight into the perceptions of building workers held by the salaried members of the industry. Here the relative social status of all those contributing to the building process is compared, the architect, being perceived as having the highest status followed closely by the other construction professions and the operatives having the lowest. The most significant finding, in relation to this study, was that where the participants are rated by 'contribution to the building process', the building operative was considered to make the lowest contribution of those directly involved with the building process on site while the contractor was seen as contributing the most (Figure 10.1). Although based only on a small sample of 97 respondents, none of them operatives, the findings are a sad indictment of industry attitudes towards building workers and the lack of recognition of their skills in the 1960s.

Conclusion

> [T]he architect should design with a clear idea of what the factory is best able to make even if this entails spending half his designing time there. But the builder is the person who has to assemble the components on the site – perhaps 60 feet off the ground – so surely he must have a say in their design too.
>
> Stirrat Johnson-Marshall, 1950[30]

Stirrat Johnson-Marshall, architect and 'inspirer, enabler and eminence of English post-war school building' held far-sighted ideas on the construction process.[31] This he considered should be a collaborative effort depending on long-term relationships between designers, researchers, builders, manufacturers and users. Unlike late twentieth-century, and still current,

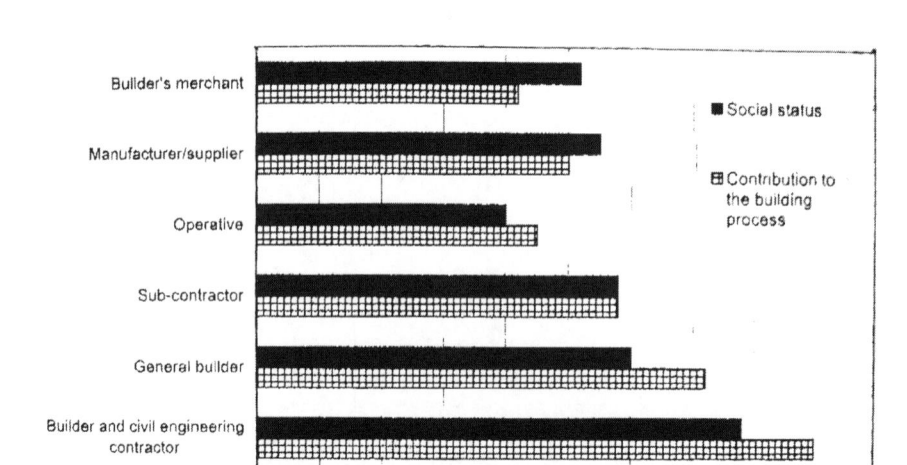

180 *Architects and building workers*

Figure 10.1 Ratings of 'contribution to the building process' and 'social status' by occupation

Source: Redrawn from Higgins and Jessop's *Communications in the Building Industry: Pilot Study*, London: Tavistock Institute, 1963.

ideas on 'partnering', where relationships between contract partners are emphasised to the general exclusion of site activities. That architects should not only consider the conditions under which their creations are built but suggest that the builder have an input into their design is now almost inconceivable. The quote by Johnson-Marshall opening this chapter was made at a historical moment when for the only time in the twentieth century there were a substantial proportion of architects either employed by central and local government or, if in private practice, engaged on publicly-funded schemes. They were employees, in the same sense as building workers were direct employees of large contractors or local authorities. This fact alone, although certainly contributing, did not engender solidarity between these two groups – but a commitment to the ideals of the Welfare State on the part of a significant number of architects and the building trade unions provided some common ground for communication.

The reconstruction of Britain after the ravages of war, the building of new schools, hospitals, housing, and the renewal of the infrastructure of transport systems and power stations, was the overriding concern of early post-war policy for the built environment under the Labour government of 1945–51. In many cases, policy decisions were based on a large amount of research anticipating reconstruction, which had been carried out during the war under the aegis of an extensive network of committees reporting to various government departments. All of it predicted major shortages of both materials and skilled labour and the solutions, for the building industry, were labour-saving techniques, new materials and more efficient organisation. Together with prefabrication, of components and entire dwelling units, standardisation and the architectural, modernist concept of rationalisation, this list of hoped-for improvements came to be described by the shorthand term 'industrialisation'.

Johnson-Marshall's quote also encapsulates the key themes repeated in the rhetoric associated with the concept of industrialisation: factory-based mass-production of components and their subsequent assembly at the building site. Even as industrialisation was largely accepted as inevitable in order to rebuild Britain, the ways in which it should be implemented were argued over at length and in great detail in the architectural and building press and in conferences and meetings over the ensuing decades. As well as an intense interest in the technical building process, many architects also felt a moral commitment to reconstruction and the Welfare State. As Andrew Derbyshire reflected many years later:

> And the idea of the Welfare State was something that we all signed up to. It was something to do with rationing in the hardship of the war, and the sacrifices we'd all made, and building homes for heroes, you know, seemed the right thing to do! There was a moral atmosphere of doing good, doing the right thing, service, service to the people. That was very strong, and nobody questioned it. I never questioned it. I'm a lifelong socialist. I believe in social justice. That has certainly affected everything I've done, and all the reasons why I've done it have been imbued with that feeling, and I think it was very common in my generation. You'll find, if you talk to my old colleagues now, they'll bemoan the passing of that particular ethos.[32]

Commitment to the Welfare State, social justice and building a better post-war world was also central to the position taken by the building trade unions. Luke Fawcett, General Secretary of the AUBTW, echoing the sentiments of Ruskin, reminded workers that 'No work is so socially necessary as building construction', and exhorted them ' . . . to build the towns and cities of the future so that in worth and ordered architectural dignity and beauty they will surpass those built in the past'.[33]

Many building workers were indeed striving to create a better world for the next generation and the political struggles of workers on the New Towns, particularly Stevenage are testament to this.[34] However, the manner in which reconstruction was to be achieved threatened the status and livelihoods of many of those charged with the task, as the particular form of industrialisation promoted by the state was underpinned by a Taylorist rationale that aimed to remove as much 'skilled', at the time understood as 'craft', work as possible away from the building site and into the factory and especially to reduce input from the 'wet trades' – mainly bricklayers. Some commentators have suggested that the building unions were forced to agree to the introduction of new methods of building because to oppose them would be seen as also opposing the ideals of the Welfare State (Finnimore, 1989: 114) but that would be to simplify the debates held between the craft and general building workers unions. Many members followed closely developments in the Soviet Union, where industrialisation was being enthusiastically promoted and were well informed about new processes and emerging skills. The main craft unions were less likely to support industrialisation, seeing it as an attack on traditional skills, but for the general unions, it opened up opportunities for new members from the new occupations associated with industrialisation, and raised anew the potential to form one industrial union for all building workers.

Similar polarities can be seen in the architectural profession, the RIBA opposing industrialisation, fearing a reduction in the role of the architect while large numbers of individual architects embraced it wholeheartedly as the expression of a rational, functional architecture entirely appropriate for the social betterment of the post-war world. The activities of the architect-dominated Modular Society are typical of those architects promoting industrialisation through the use of dimensional co-ordination in the belief it would ensure architects remained in control of design with the labour force regarded as an element of the technical process (see Figure 10.2). Few architects gave building workers much consideration but those who did, and contributed to this book, were committed modernists and remained involved with the state-promoted drive to industrialise the construction industry, while at the same time attempting direct engagement with workers on site.

The 1960s also saw increasing distance between trade union officials and rank and file activists exemplified in the bitter dispute at the Barbican development, resulting in a government enquiry (McGuire *et al.*, 2013). In 1970, after many amalgamations, a single construction union, UCATT, was founded, paving the way for a short-lived period of union strength and providing a new voice for labour in the industry. By the early 1970s a new generation of architects were questioning the Establishment view of professionalism espoused by the RIBA. For some of these architects, active in the New Architecture Movement, feminist practices like Matrix, and in building co-operatives, the contribution of building workers became, once again, a visible and valued part of the building process.

Figure 10.2 A75 Metric building under construction
Source: *Modular Quarterly*, 1967, No. 2.

Appendix 1

Table A.1 Visits by Modular Society members to modular buildings, 1953–56

Location	Date	Type of Modularity	No. of attendees	Guides
Summerswood School, Borehamwood, Herts.	June 1953	Planning grid and structural grid of 40"	90	W. Lacey B. Martin A. Williams
Wokingham School, Berks.	July 1953	Planning grid 40"	45	M.W. Lee G.B Oddie David Medd J.K.L. Kitchin
Dartford Technical College, Kent	June 1954	Planning grid 50"	N/k	J.H. Garnham Wright
Oakridge Farm Infants School, Basingstoke	July 1954	Bristol aeroplane Mark 1A system	20	Richard Sheppard Jean Sheppard Geoffrey Robson
Worthing Secondary Technical School	September 1954	Planning grid 40"	16	N/k
Arcon Prototype site, Taylor-Woodrow's Works, Greenford, Middsx.	September 1954	Multiple grids for different components	35	A.M. Gear
Prendergast County Complement School, Catford	October 1954	Planning grid 40", vertical grid 8"	40	J.L. Martin G.F. Horsfall W.J. Smith

Housing at Basingstoke	October 1954	Bellrock panels 2′ wide × 4″ thick	16	Eric Almond
Unity Structures at Ideal Home Exhibition	March 1955	3′ 0″ grid	61	G.K. Findley
Laboratory for Arthur Guinness in Seco Mk VIII System	April 1955	40″ constructional grid	41	N/k
Marchwood Generating Station, Southampton	August 1955	13′ planning grid, $9^{3}/_{4}$″ vertical & horizontal module	34	Frankland Dark Noel Hunter
Service Stations: i) Aylesbury for Shell-mex ii) Reading for British Petroleum	August 1955	4″/40″ planning grid, 4″ module	40	D.A. Birchett (Company architect for Shell-mex) M.A. Wolstenholme of Wallis, Gilbert & Partners
Cole Green Lane, Secondary modern school, Welwyn Garden City	19 April 1956	Planning grid 40″, structural grid 13′ 4″	31	William Tatton-Brown (written report)
Bowater Paper Corporation Ltd.	22 May 1956	Planning grid 40″, vertical module 10″	23	J.T. Pinion and R.L. Brewerton of Farmer and Dark Architects.
Shell Offices, Shell Haven, Essex	20 June 1956	Planning grid 40″	30	Howard Lobb John Ratcliffe

Source: *Modular Society Transactions* and *MQ* 1953–1956.

Appendix 2
Letter from Mark Hartland Thomas to two members of the Modular society on RIBA endorsement of four-inch module

26th May '62

...

Many thanks for yours of the 24th giving me the RIBA resolution. I am very glad to have this information.

It is the consummation of a movement that I started at the RIBA in 1946 at the ASB and sidestepped to the BSI later in the same year to avoid being blocked by opposition in the RIBA Council. Seven years later, in 1953, when both BSI and BRS had declined to take up the subject intensively, I started the Modular Society. That was the signal for all the others find an interest in the subject, except the RIBA, but all through these years the Modular Society has maintained the lead in initiative in development and in technical expertise: I do not except even EPA 174 from this claim.

Now if the RIBA is to swing into action in the way that you outline, and I welcome it, we must take a cold hard look at the position of the Modular Society. We must not make the same mistake as the DIA which should have gracefully wound itself up when COID took over. The reiterated admonition in the RIBA Report about the architect assuming control of industrialised building is exactly what we were saying in 1946 from the ASB to the unheeding ears on the RIBA Council. Now at last their successors have listened and thanks to the efforts of a small number of RIBA members who are also members of the Modular Society the line has held and the slow moving council does not find the architect's position already lost, as it did in town planning.

The issue is now quite simply this: either the RIBA adopts the Modular Society as its chosen instrument (right through your provisional programme of publicity, starting with the RIBAJ leader) or we wind up the Modular Society, announcing that the takeover by the RIBA has brought our work to completion in 10 years.

What I am not willing to contemplate is any sort of half-way house in which the RIBA steals the wind from all our sails so that our numbers and potential members say that they get all the guidance they need from the RIBA and have no reason to subscribe to the Modular Society.

The Modular Society is hanging on by a shoe-string diminishing in length. To meet the new situation it must have a massive increase in membership, or wind up.

The USA parallel is interesting. There, the modular movement started as a department of the AIA and was run from the Institute for many years until it petered out. It was revived, after a gap, by the foundation of the Modular Building Standards Association in 1958 in direct imitation of our Modular Society, for they consulted me and borrowed a copy of our Memorandum and Articles.

This may mean that if the RIBA squeezes out the Modular Society, it will find after a while that it must recreate something similar in order to bring into partnership the rest of the building industry and its professions.

I shall be raising the matter at the Council of the Modular Society at its meeting on the 31st. I hope you will be present. We shall of course respect the embargo, though I can see little reason for it and am profoundly suspicious. The embargo could allow opposition to be whipped up before the announcement and may have been engineered for that very purpose.

Yours sincerely,

Mark

Source: Bruce Martin's papers. Letter on Modular Society notepaper.

Glossary of modular terms

Basic module – M	A module with the size of 4 inches or 100 mm used to co-ordinate the dimensions of components and of buildings.
Closed system	A building system of industrially manufactured components that dimensionally co-ordinate with each other.
Component	An industrial product manufactured as an independent unit possessing fixed dimensions in at least two dimensions.
Dimensional co-ordination	The establishment of a range of related dimensions for common use in sizing buildings and the components which make up those buildings.
Grid	A reference system of lines or planes which may be used on a drawing or on the site. The lines are straight and generally at right angles and the grid is often square and at one standard spacing, but tartan grids with two or more different intervals between grid lines are also common.
Keeping station	Components which keep station are those which are placed within the spaces allocated to them on a reference grid.
Modular component	A building component whose co-ordinating dimensions have modular sizes.
Modular co-ordination	A method of sizing the dimensions of building components and of building on the basis of a basic module.
Modular grid	A grid in which the intervals between lines is in multiples of the basic module.
Module	A convenient size which is used as an increment.
Open system	A system of industrially manufactured components that are, theoretically, dimensionally related and universally interchangeable.

Source: UN (1974)

Notes

Introduction

1 Marian Bowley, *The British Building Industry: Four Studies in Response and Resistance to Change* (Cambridge: Cambridge University Press, 1966), pp. 438–439.
2 David Gann, *Building Innovation* (London: Thomas Telford, 2000), p. 10.
3 David Edgerton, *Science, Technology and the British Industrial 'Decline' 1870–1970* (Cambridge: Cambridge University Press, 1996).
4 Corelli Barnett, *The Audit of War: The Illusion and Reality of Britain as a Great Nation* (London: Macmillan, 1986); Martin Wiener, *English Culture and the Decline of the Industrial Spirit* (Cambridge: Cambridge University Press, 1981).
5 David Landes, *The Wealth and Poverty of Nations* (London: Little Brown and Company, 1998), p. 458.
6 Michael Ball, *Rebuilding Construction: Economic Change in the British Construction Industry* (London: Routledge, 1988).
7 Ibid., p. 74.
8 Ibid., p. 111.
9 The summary of historical sociological theories used here is found in full in P. Burke, *History and Social Theory* (Cambridge: Cambridge University Press, 1992), Chapter 5, 'Social Theory and Social Change', pp. 131–165.
10 *Bartlett International Summer School Proceedings (1979–92)* (London: University College London), vols 1–14.
11 Building Research Station, *Prefabrication: A History of its Development in Britain*, ed. R.B. White (London: HMSO, 1965), p. 300.
12 Ibid., p. 302.
13 Patrick Dunleavy, *The Politics of Mass Housing in Britain 1945–75: A Study of Corporate Power and Professional Influence in the Welfare State* (Oxford: Clarendon Press, 1981). See Chapter 3.
14 The density guidelines published by the MHLG in 1958 came under attack in the 1960s in Peter Stone's *Housing, Town Development and Land Costs* (London: Estates Gazette, 1963). The first major theoretical work arguing against high density through high-rise was L. Martin and L. March, *Urban Space and Structures* (Cambridge: Cambridge University Press, 1972).
15 Brian Finnimore, *Houses from the Factory: System Building and the Welfare State* (London: Rivers Oram Press, 1989). See pp. 111–120.
16 Miles Glendinning and Stephan Muthesius, *Tower Block: Modern Public Housing in England, Scotland, Wales, and Northern Ireland* (New Haven, CT: Yale University Press, 1994).
17 Ibid., p. 3.

18 Ibid., pp. 215–217, 318.
19 Ibid., p. 195.
20 Ibid., p. 74.
21 Barry Russell, *Building Systems, Industrialization and Architecture* (London: John Wiley & Sons, Ltd., 1981).
22 Linda Clarke, *Building Capitalism: Historical Change and the Labour Process in the Production of the Built Environment* (London: Routledge, 1992).
23 Linda Clarke and C. Winch C. (2006) 'A European Skills Framework? – But What Are Skills? Anglo-Saxon Versus German Concepts', *Journal of Education and Work* 19(3): 255–269.
24 John Ruskin, *The Seven Lamps of Architecture, The Lamp of Life XXIV*, (London: George Routledge & Sons Ltd., 1910) p. 181.
25 John Ruskin, *Fors Clavigera. Letters to the Workmen and Labourers of Great Britain*, ed. Dinah Birch (Edinburgh: Edinburgh University Press, 2000).
26 John Ruskin, *Selected Writings*, ed. Dinah Birch, *The Stones of Venice*, II, 1853, *The Nature of Gothic* (Oxford: Oxford University Press, 2004), p. 47.
27 Ibid.
28 Schinkel, quoted in Brian Hanson, *Architecs and the 'Building World' from Chambers to Ruskin: Constructing Authority* (Cambridge: Cambridge University Press, 2003).
29 John Unrau, 'Ruskin, the Workman and the Savageness of Gothic', in R. Hewison (ed.) *New Approaches to Ruskin* (London: Routledge, 1981), pp. 33–50.
30 John Ruskin, 'On the Nature of Gothic Architecture: and Herein of the True Functions of the Workman in Art'. Reprinted from the sixth chapter of the second volume of Mr. Ruskin's *Stones of Venice* (London: G. Allen, 1892).
31 William Morris, *Political Writings of William Morris*, ed. A.L. Morton (London: Lawrence & Wishart, 1973).
32 H. Braverman, *Labor and Monopoly Capital: The Degradation of Work in the Twentieth Century* (New York: Monthly Review Press, 1974), p. 443.
33 L. Clarke, 'The Changing Structure and Significance of Apprenticeship with Special Reference to Construction', in P. Ainley and H. Rainbird (eds) *Apprenticeship* (London: Kogan Page, 1999).
34 R. Sennett, *The Craftsman* (London: Allen Lane, 2008).
35 R. Biernacki, *The Fabrication of Labor: Germany and Britain 1640–1914*, (Berkeley, CA: University of California Press, 1995).
36 M. Brockmann, L. Clarke and C. Winch, 'Competence-Based Vocational Education and Training (VET) in Europe: The Cases of England and France', *Vocations and Learning* 1(3) (2008): 227–244.
37 Biernacki op. cit., p. 83.
38 Ibid., Chapter 3, 'The Control of Time and Space', p. 141.
39 See S. Prais and H. Steedman, 'Vocational Training in France and Britain: The Building Trades', *National Institute of Economic and Social Research (NIESR) Review* 116 (May 1986); and A. Green, 'Poor Relation or "The General Education of the Working Class": VET in a Historical Perspective,' paper presented at ESRC seminar series; 'Vocational Education: Historical Developments, Aims and Values' (2002).

1 The industrialisation of building

1 Hannes Meyer, writing in 1928, in Claude Schnaidt, *Hannes Meyer: Buildings, Projects and Writings* (London: A. Tiranti, 1965), p. 94.
2 R.T. Walters, 'Towards Industrialised Building', *RIBAJ* (February 1957): 150–153.
3 Ministry of Works, *Machines for the Modern Builder* (London: HMSO, 1950c).

The *AJ* featured a regular report on 'The Industry', edited by Philip Scholberg with occasional special features, for example, 'He Doubles the Speed of Bricklaying', *AJ* (29 May 1947). Eric de Maré edited a well-illustrated book aimed at architects called *New Ways of Building* (London: Architectural Press, 1948), which included contributions from Cecil Handisyde and Philip Scholberg.
4 For example, D. Preziosi, *Minoan Architectural Design: Formation and Significance* (Berlin: Mouton, 1983); G. Herbert, *Pioneers of Prefabrication: The British Contribution in the Nineteenth Century* (Baltimore, MD: Johns Hopkins University Press, 1978); A. Hoagland, 'The Invariable Model. Standardisation and Military Architecture in Wyoming 1860–1900', *Journal of the Society of Architectural Historians* 57(3) (1998): 298–315.
5 *Tudor Walters Report*, Part XI Organisation (London: HMSO, 1918), pp. 74–78.
6 Department of Health, *New Methods of House Construction: Interim Report* (London: HMSO, 1924).
7 F.J. Schwartz, *The Werkbund and Design Theory: Mass Culture Before the First World War* (New Haven, CT: Yale University Press, 1996).
8 Peter Meyer, 'Typisierung und Normung', in *Moderne Architektur und Tradition* (Zurich, 1928), quoted in Schwartz (1996).
9 The Albert Farwell Bemis Foundation was founded in 1938. Based at the Massachusetts Institute of Technology (MIT), it became an influential centre for research into prefabrication in housing with strong international connections. In 1940, the director, John Burchard, arranged for Alvar Aalto to become visiting professor.
10 Alfred C. Bossom, *Building to the Skies: The Romance of the Skyscraper* (London: Studio, 1934).
11 Ibid., p. 151.
12 See Andrew Saint, *Towards a Social Architecture: The Role of School-Building in Post-War England* (New Haven, CT: Yale University Press, 1987), p. 25. Among them were Guy Oddie, W.A. Allen and Cecil Handisyde.
13 For example, *Design for Britain*, edited by Edwin Fairchild, was a series of 36 pamphlets published by the Co-operative Permanent Building Society. It included No. 27, *Post-War Building* by Alfred Bossom. In a similar vein, the *Rebuilding Britain* series, edited by F.J. Osborn, included No. 10, *Civic Design and the Home*, 1943, by Arnold Whittick, with a Foreword by Patrick Abercrombie, and No. 5 *Plan for Living: The Architect's Part*, 1941, by Clough Williams-Ellis.
14 BRS, *Prefabrication: A History of its Development in Britain*, ed. R.B. White (London: HMSO, 1965), p. 122.
15 C.M. Kohan, *Works and Buildings* (London: HMSO, 1952), p. 142.
16 Ministry of Works, *Training for the Building Industry* (London: HMSO, 1943).
17 Ministry of Works, *Working Party Report on Building* (London: HMSO, 1950).
18 Conservative Party policy on housing is examined by Harriet Jones in '"This is Magnificent": 300,000 Houses a Year and the Tory Revival after 1945', *Contemporary British History*, 14(1) (Spring 2000): 99–121.
19 Author's interview with Roger Walters, April 1999.
20 *IBSAC* (1963) 1(1): 10.
21 A.V. Knight, 'A Critique of Work Study', *Work Study and Industrial Engineering* 3 (1956): 81.
22 D.G.R. Bonnell, 'Operational Research in Building', in *Building Research and Documentation: The First CIB Congress* (Rotterdam: CIB 1959), pp. 486–493.
23 MHLG, *Industrialised Housebuilding* Circular No. 76/65 (London: HMSO, 1965).

24 Ministry of Housing and Local Government, *Housing Programme 1965–70* (London: HMSO, 1965).
25 MHLG Circular No.76/65, op. cit.
26 T.P. Bennett, 'The Architect and the Organisation of Post-War Building', *RIBAJ* (Jan. 1943): 63–74.
27 Misha Black, *The Architect's Anguish*. Text of the annual Harry Hardy Peach Memorial Lecture (15 Feb. 1962) by the Professor of Industrial Design (Engineering) (Leicester: Leicester University Press, 1962).
28 A 1970 UN publication defined modular co-ordination as 'a method of sizing the dimensions of building components on the basis of a single module' and dimensional co-ordination as 'the establishment of a range of related dimensions for common use in sizing buildings and components which make up those buildings'.
29 Mark Hartland Thomas, 'Cheaper Building: The Contribution of Modular Co-ordination', *Journal of the Royal Society of Arts* 101 (1953): 98–101.
30 A.W. Cleeve Barr, *Public Authority Housing* (London: Batsford, 1958), p. 120.
31 R.W. Postgate, *The Builder's History* (London: NFBTO, 1923).
32 F. Zweig, *Productivity and the Trade Unions* (Oxford: Basil Blackwell, 1951), pp. 8 and 35.
33 MRC MSS. 78/BO/UM/4/1/21, *NFBTO Conference 1950 Report* p. 155.
34 MRC MSS. 78/BO/UM/4/1/23, *NFBTO Conference 1952 Report* p. 109.
35 H.J.O. Weaver, 'Innovation and the Operatives', in *Supplement to Innovation in Building: Contributions to the CIB Congress, Cambridge* (Amsterdam: Elsevier, 1962, pp. 64–71.
36 Author's interview with Donald Bishop. Bishop became friendly with Harry Weaver while researching bricklayers' productivity at the BRS.
37 Ibid.
38 MRC MSS. 78/BO/UM/4/2/3. April 1959, Report of Conference on New Techniques in the Building Trades.
39 MRC MSS. 78/BO/UM/4/1/33. NFBTO Annual Conference Report 1962, p. 117.
40 MRC MSS. 78/BO/UM/4/1/33. NFBTO Annual Conference Report 1962, p. 118.
41 MRC MSS. 78/BO/UM/4/1/123. pp. 110–111.
42 *IBSAC* 1(1) (1963): 12.

2 The building industry during war and reconstruction

1 Between 1942 and 1943, the Ministry of Works and Buildings took over responsibility for planning from the Ministry of Health and was renamed the Ministry of Works and Planning. With the setting up of the Ministry of Town and Country Planning in 1943, it was again renamed as the Ministry of Works. In 1962, the department was reconstituted as the Ministry of Public Building and Works until 1970, when it merged with the Ministry of Transport and the Ministry of Housing and Local Government to form the Department of the Environment.
2 C.M. Kohan, *Works and Buildings*, History of the Second World War, United Kingdom Civil Series, edited by W.K. Hancock (London: HMSO, 1952), p. 78.
3 Ibid., pp. 39–40.
4 Leslie Wood, *A Union to Build: The Story of UCATT* (London: Lawrence & Wishart, 1979), p. 64.
5 BRS, *Collection of Construction Statistics* (1971), Table 7.19.
6 TNA PRO LAB 8/518. *The Post-War Demand for Labour in the Building and Civil Engineering Industries*, G.D.H. Cole, p. 12.
7 Ministry of Works Statistics. The category 'Skilled craftsmen' includes: carpenters and joiners, bricklayers, plumbers, glaziers, masons, and electricians.

Notes 193

8 Leslie Wallis, *The Building Industry: Its Work and Organisation* (London: J. M. Dent & Sons, 1945). This was the private sector view from a small builder who was also a J.P. and President of Maidstone Rotary Club. He became President of the BINC between 1943 and 1944.
9 E.D. Simon, *Rebuilding Britain: A Twenty-year Plan* (London: Gollancz, 1945), passage quoted in Kohan, 1952, op. cit., p. 8.
10 TNA PRO LAB 8/867. *Designated Craftsmen Scheme.* Letter from T.A. Barlow, Treasury Chamber, to Hugh Beaver, Director-General, Ministry of Works.
11 TNA PRO LAB 8/867. Letter to Gould at the Ministry of Labour and National Service.
12 TNA PRO LAB 8/867.
13 TNA PRO LAB 8/867. Letter to Voysey, Ministry of Works from London Master Builder's Association, 30 July 1943.
14 National Archives CAB 21/1527, Production Council, *The Building Industry and the War 1941–3.*
15 TUC LRD 2/B/1 *New Builder's Leader*, 6(6) (May 1941): 24.
16 TUC HD 6661 W8 *Amalgamated Society of Woodworkers Monthly Journal*, 2 (Feb. 1942): 83.
17 Op. cit. p. 85.
18 TUC LRD 2/B/Building *New Builder's Leader and Electricians Journal*, 8(1943): 8.
19 TUC HD 6661 NFBTO *The Operative Builder, 1953–60*, B9.
20 TNA PRO LAB 8/464. *Memorandum of War Time Agreement on Employment of Women in the Building Industry 1941–42.*
21 D. Hall, *Cornerstone: A Study of Britain's Building Industry* (London: Lawrence & Wishart, 1948), p. 82.
22 See *NBL and EJ*, 'The Building Trade Worker and D-Day'; 10(4) (January 1945): 28.
23 C. Briar, *Working for Women: Gendered Work and Welfare Policies in Twentieth Century Britain* (London: UCL Press, 1997), p. 84.
24 TUC LRD 2/B/1/Building *New Builder's Leader*, January 1945 (10)4.
25 TUC HD 9715/6 NFBTO Annual Conference, 1945.
26 See Linda Clarke and Christine Wall, '"A Woman's Place Is Where She Wants to Work": Barriers to the Retention of Women in the Building Industry after the Second World War', *Scottish Labour History* (2010): 16–39.
27 Laings were already in a strong position to benefit from post-war house building through developing their *in-situ* concrete system, the Easiform house, helped by expertise of the Building Research Station (BRS) under the Directorate of Post-War Building set up by Ministry of Works in 1942.
28 G.D.H. Cole, 'The Building Industry after the War', *Agenda. A Quarterly Journal of Reconstruction* 2 (1943): 148. Oxford: London School of Economics and Political Science.
29 Anon, *Britain Without Capitalists. A Study of What Industry in a Soviet Britain Could Achieve, by a Group of Economists, Scientists and Technicians* (London: Lawrence & Wishart, 1936), p. 140.
30 Ibid., Chapter IV, *Building*, was written by J.D. Bernal.
31 G.D.H. Cole, *Building and Planning* (London: Cassell, 1945), p. 246.
32 Harry Barham, *Building as a Public Service* (Dunfermline, J.B. Mackie, 1945).
33 Ibid., p. 14.
34 Ibid., p. 3.
35 Harry Barham, *The Building Industry: A Criticism and Plan for the Future*, Industrial Democracy Series, No. 1 (London: St. Botolph Publishing Ltd., 1947).
36 Hall (1948), op. cit., p. 133.

194 Notes

37 MRC, *The Operative Builder* 10(5) (Sept. 1957): 315–316.
38 Ibid., p. 316.
39 W.S. Hilton, *Foes to Tyranny: A History of the AUBTW* (London: AUBTW, 1963), p. 266.
40 Ibid., p. 267.
41 Richard Coppock and Harry Heumann, *Design for Labour*, No. 13 in Design for Britain Series (London and Letchworth: J.M. Dent & Sons, 1948).
42 *Nationalisation of the Building Industry*. Scheme Presented to the 33rd Conference of the NFBTO in Ayr, June 1950, NFBTO.
43 *Building as a Public Service* (National Federation of Building Trade Operatives, 1956), p. 10.
44 Ibid., p. 6.
45 Anthony Jackson, *The Politics of Architecture: A History of Modern Architecture in Britain* (London: Architectural Press, 1970), p. 78. An excellent and detailed account of the design work carried out by architects involved in the Second World War is found in *Architecture in Uniform: Designing and Building for the Second World War* by Jean-Louis Cohen, Canadian Centre for Architecture (Paris: Hazan, 2011).
46 Jackson (1970), op. cit., p. 79.
47 Kohan (1952), op. cit., pp. 387–388.
48 Described in Cohen (2011), op. cit., pp. 164–170.
49 *Architectural Design* 10(10) (1940): 232–235.
50 Author's interview with Alan Crocker, architect and author, in 2003.
51 Author's interview with Alan Meikle, 2000.
52 TNA PRO LAB 8/1095. *Post-war Reconstruction: Control of Numbers of Architects Needed*.
53 TNA PRO LAB 8/1095. Letter from E.G. McAlpine, Ministry of Education to Miss B. Grainger, Ministry of Labour and National Service, 5 Feb. 1947.
54 Ibid.
55 *RIBAJ* 57(2) (December 1949): 59–60.
56 *AJ* (December 15 1949): 66.
57 TNA PRO LAB 8/424. Quoted in *Notes from a Meeting of the Hankey Technical Personnel Committee* (undated).
58 TNA PRO LAB 8/518. The Education Committee consisted of: E.D. Simon (chairman), G. Burt, R. Coppock, L. Fawcett, Hindley, Mr. Laing, Loudon, McTaggart, Thomas, Pitt, G.D.H. Cole and representatives of various government departments.
59 TNA PRO LAB 8/518. Advance Copy of Part III of G.D.H. Cole's Report, p. 4.
60 Ibid., p. 13.
61 TNA PRO LAB 8/1095. Central Council for Works and Buildings, Training Sub Committee Minutes, 31 May 1943.
62 TNA PRO LAB 8/1095. ibid.
63 TNA PRO LAB 8/518. *The Post-War Demand for Labour in the Building and Civil Engineering Industries* by G.D.H. Cole, and the Simon Committee's *Report on Training for the Building Industry* (London: HMSO, 1942).
64 *Report on Training for the Building Industry* (London: HMSO, 1943), pts 44 and 45.
65 Ibid., pt 54.
66 Cmd. 6428.
67 Ibid., pt 13.
68 G.D.H. Cole, 'The Labour Problem in the Building Industry', *Agenda: A Quarterly Journal of Reconstruction* 1 (1942): 129–143.
69 G.D.H. Cole, 'The Building Industry after the War', *Agenda. A Quarterly Journal of Reconstruction* 2 (1943): 146–152.
70 Ibid., p. 152.

3 Education and training

1. Alan Powers, 'Arts and Crafts to Monumental Classicism: The Institutionalisation of Architectural Classicism 1900–1914', in Neil Bingham (ed.) *The Education of the Architect, Proceedings of the 22nd Annual Symposium of the Society of Architectural Historians of Great Britain* (London: 1993), pp. 34–38.
2. RIBA, *Report of the Special Committee on Architectural Education* (London: RIBA, 1943), p. 7.
3. *RIBAJ* 63(4) (February 1956): 165.
4. Mark Crinson and Jules Lubbock, *Architecture, Art or Profession? Three Hundred Years of Architectural Education in Britain* (London: Prince of Wales Institute of Architecture, 1994). See Chapter 3, 'The Modernist Academy 1938–60'.
5. TNA PRO ED 150/45. Letter from Michael Waterhouse, President of the RIBA to G. Tomlinson, Minister of Education, 13 January 1950.
6. TNA PRO ED 150/45. Minute Sheet, 18 January 1950.
7. Ibid.
8. A. Saint, *Towards a Social Architecture: The Role of School-Building in Post-War England* (New Haven, CT: Yale University Press, 1987), p. 245.
9. 'Conference on Architectural Education: Report by Leslie Martin', *RIBAJ* 65(8) (1958): 279–282.
10. Denis Harper 'Some Notes on the Education of the Architect and his Technical Assistant', *RIBAJ* 71(10) (1964): 441–442.
11. Ibid., p. 442.
12. Elizabeth Layton, *Report on the Practical Training of Architects* (London: RIBA, 1962), pp. 11 and 25.
13. Ibid., p. 44.
14. Conference on Building Training where the three main speakers were: D.E. Woodbine Parish, W. James and D.H. MacMorran, *RIBAJ* 63(4) (February 1956).
15. Noel Hall, *Report of the Joint Committee on Training in the Building Industry* (London: RIBA, 1965).
16. Ibid., p. 5.
17. Ibid., p. 8.
18. *RIBAJ* 72(2) (February 1965): 54. The Noel Hall recommendations were summarised on pp. 56–57 of the same issue under the heading: 'Agreement on Joint Training: Urgent Need for Fresh Approach'.
19. R. Llewelyn-Davies and J. Weeks, 'Educating for Building', Paper Number 3, *BASA Second Report on Architectural Education, Building for People* (London: BASA, 1962).
20. Ibid., p. 68.
21. *A Course of Education for the Architect: A Report of a Study Group at the Department of Architecture, Surveying, Building and Interior Design* (London: The Northern Polytechnic, 1963).
22. Ibid., p. 8.
23. Curriculum change and the content of architects training over the twentieth century are described in detail by Andrew Saint in *Architect and Engineer: A Study in Sibling Rivalry* (London: Yale University Press, 2007), pp. 462–482.
24. *The Architect and Building News* 195(4201) (24 June 1949): 3.
25. MRC MSS. 78/BO/UM/4/1/25. *NFBTO Annual Conference 1954 Report*.
26. MRC MSS. 78/BO/UM/4/1/28. *NFBTO Annual Conference 1957 Report*, Comment by F.E. Shrosbee, General Secretary of the ABT.
27. Preface to *BATC Final Report* (1956) by the chairman F.W. Leggett.
28. *The Builder* 176(5525) (7 January 1949): 7.

29 MRC MSS. 78/BO/UM/4/1/21. *NFBTO Annual Conference 1950*. Richard Coppock's reference to architects, p. 57.

4 Post-war change: management and organisation

1 Department of Employment and Productivity, *Report of the Committee of Inquiry under Professor E.H. Phelps-Brown into Certain Matters Concerning Labour in Building and Civil Engineering* (London: HMSO, 1968), p. 95.
2 Peter Hennessy, *Never Again. Britain 1945–51* (London: Vintage, 1993), Chapter 9, p. 359, and Alan Sked and Chris Cook, *Post-war Britain: A Political History* (Harmondsworth: Penguin, 1984), pp. 34–35.
3 M. Francis, *Ideas and Policies under Labour, 1945–51: Building a New Britain* (Manchester: Manchester University Press, 1997), p. 43.
4 J. Tomlinson, 'The Failure of the Anglo-American Council on Productivity', *Business History* 33 (1991): 82–92.
5 Ibid., p. 83.
6 A full account of the trade unions and changing industrial relations in relation to the AACP and other aspects of the productivity drive is given in A. Carew, *Labour under the Marshall Plan: The Politics of Productivity and the Marketing of Management Science* (Manchester: Manchester University Press, 1987).
7 Tomlinson, 'The Failure of the AACP', (1991) op. cit.
8 British Productivity Council, *AACP Report on Building* (London: ACCP, 1950).
9 J. Tomlinson, 'The Politics of Economic Measurement: The "Productivity Problem" in the 1940s', in A. Hopwood, and P. Miller (eds) *Accounting as Social and Institutional Practice* (Cambridge: Cambridge University Press, 1994). See also J. Tomlinson, 'The Labour Government and the Trade Unions, 1945–51', in N. Tiratsoo (ed.) *The Attlee Years* (London: Pinter, 1991).
10 Ministry of Works Report, *Payment by Results in Building and Civil Engineering* (London: HMSO, 1947a) and I. Bowen (Chief Statistical Officer for the Ministry of Works), 'The Control of Building', in *Lessons of the British War Economy* (Cambridge: Cambridge University Press, 1951).
11 Ministry of Works, *Working Party Report on Building* (London: HMSO, 1950b), p. 11.
12 Ministry of Health, *The Cost of House-Building* (London: HMSO, 1948).
13 AACP (1950), op. cit., p. 2.
14 Carew, op. cit. (1989), p. 139. In this case, they were Patrick Bates (plumber), Francis Beazley (administrative assistant at the NFBTO), William Clarke (plasterer and branch secretary, NAOP), William Johnstone (carpenter employed by Dove Bros. Ltd.), Glynn Lloyd (painter and decorator and Apprentice Master Scheme instructor) and John McKechnie (Clerk of Works and member of Executive Council, AUBTW).
15 AACP (1950), op. cit., p.21.
16 Ibid., para. 26, p. 22.
17 A quote from Alistair Cooke's *Letter from America* BBC Radio Programme.
18 Ibid, para. 26, p. 14.
19 Ibid., p. 65.
20 Carew (1995), op. cit., p. 145. The other industry unions that publicly disagreed with their respective AACP Reports were printing, foundry and hosiery.
21 MRC. MSS.78/BO/UM/43. *Operative Builder* 3(4) (1950): 142.
22 British Productivity Council, *Getting Together: A Review of Productivity in the Building Industry* (London: BPC, 1954), p. 6.
23 The Working Party was chaired by Sir Thomas Phillips and comprised 16 members including Coppock, Fawcett, Armstrong and Sandercock representing the operatives, five employers, one economist, two accountants, two from the RICS, one from the ICE and Sir Launcelot Keay representing the RIBA.

24 TNA PRO CAB/21 2026. Minutes of the Cabinet Production Committee, 31 March 1950.
25 Ministry of Works, *Working Party Report on Building* (London: HMSO, 1950b), p. 23.
26 See K.O. Morgan's *Labour in Power 1945–51* (Oxford: Oxford University Press, 1984), pp. 163–170 and Nic Bullock's *Building the Post-War World* (London: Routledge, 2002), pp. 274–275.
27 *Management Training in the Building Industry: Report of a Study Group Set Up by the British Institute of Management* (London: BIM and Board of Building Education, 1956).
28 Hansard, accessed on 26 October 2012, available at: http://hansard.millbank systems.com/commons/1950/may/22/building-industry para1683.
29 Ministry of Works, *Methods of Building in the USA* (London: HMSO, 1944).
30 Ibid., para. 68, p. 12.
31 Ibid., para. 70, p. 12.
32 International Council for Building Research, Studies and Documentation, CIB, *Building Research and Documentation: Contributions and Discussions at the First CIB Congress, Rotterdam, 1959* (Amsterdam: Elsevier, 1961).
33 Different national conceptions of skill are discussed in detail in L. Clarke and C. Winch, 'A European Skills Framework? – But What Are Skills? Anglo-Saxon Versus German Concepts', *Journal of Education and Work* 19(3) (2006): 255–269; L. Clarke and C. Winch (eds), *Vocational Education: International Approaches, Developments and Systems* (London and New York: Routledge, 2007).
34 Ministry of Works, National Building Studies, Special Report No. 18, *Productivity in House-Building (Pilot Study)* (1950); Ministry of Works, National Building Studies, Special Report No. 21, *Productivity in House-Building*, (eds) W.J. Reiners and H.F. Broughton (London: HMSO).
35 Frank Gilbreth's movement studies, together with Taylor's time and motion work, later known as work-study, were both subsumed under the broad mantle of scientific management: a strategy which emphasised the individual over teamwork and the eliciting, through payment schemes, of the maximum output of work using the maximum physical capacity of the worker. See F.W. Taylor, and S.E. Thompson, *Concrete Costs, Tables and Recommendations for Estimating the Time and Cost of Labor Operations in Concrete Construction and for Introducing Economical Methods of Management* (London: John Wiley, 1912); Frank B. Gilbreth, *Concrete System* (New York, 1908) and *Bricklaying System* (New York, 1909). The Gilbreths distanced themselves from mainstream Taylorism, considering it unethical, see, *Time Study and Motion Study as Fundamental Factors in Planning and Control: An Indictment of Stop Watch Time Study*. Read before the New York Section of The Taylor Society (December 16 1920) (The Mountainside Press), p. 16.
36 Ministry of Works, National Building Studies, Special Report No. 4, *New Methods of House Construction* (London: HMSO, 1948), p. 1.
37 Ibid.
38 Ministry of Works, National Building Studies, Report No. 10, *New Methods of House Construction. Second Report* (London: HMSO, 1949).
39 Ministry of Works, National Building Studies, Special Report No. 18, op. cit.; Ministry of Works, National Building Studies, Special Report No. 21 op. cit.; Ministry of Works, Alison Entwhistle and W.J. Reiners, NBS No. 28, *Incentives in the Building Industry* (1958); Ministry of Works, W.J. Reiners and H.F. Broughton, NBS No. 30, *A Study of Alternative Methods of House Construction* (1959); Ministry of Works, H.F. Broughton, J. Reiners and H.G. Vallings, NBS No. 31, *Mobile Tower Cranes for Two and Three Storey Building* (1960).

40 D. Bishop, 'Note on Some Factors Affecting Productivity', in Department of Employment and Productivity, *Report of the Committee of Enquiry Under Professor Phelps-Brown into Certain Matters Concerning Labour in the Building and Civil Engineering Industries* (London: HMSO, 1968), pp. 170–179.
41 Ibid., p. 177.
42 Entwhistle and Reiners (1958), op. cit.
43 Building Research Station, *Architects and Productivity*, D. Bishop, BRS Current Papers, Design Series No. 57 (London: HMSO, 1966a), Appendix, p. ii.
44 *BATC Final Report* (1956), para. 102.
45 *CITB Annual Report* (London: HMSO, 1967), p. 21.
46 'The Needs of Our Industry and the Way Ahead', Donald Gibson speaking at the 1963 RIBA Conference on The Architect and Productivity, reported in *RIBAJ* 70(9) (1963): 361–362.
47 R. A. Burgess, 'Accuracy and Productivity in IB', *Building*, 18 November 1968, pp. 189–98.
48 Ministry of Works, *Wages, Earnings and Negotiating Machinery in the Building Industry since 1886*, Chief Scientific Adviser's Division, Economics Research Section (London: HMSO, 1950d).
49 MRC MSS. 78/BO/UM/4/1/21. *NFBTO Annual Conference 1950 Report*, p. 77.
50 MRC MSS. 78/BO/UM/4/1/18-40. Reports of the NFBTO Annual Conferences.
51 MRC MSS. 78/BO/UM/4/1/28. *NFBTO Annual Conference 1957 Report*, p. 59.
52 MRC MSS. 78/BO/UM/4/1/34. *NFBTO Annual Conference 1963 Report*, p. 53.
53 MRC MSS. 78/BO/UM/4/1/34. *NFBTO Annual Conference 1963 Report*, p. 55.
54 J. Tomlinson and N. Tiratsoo, *The Conservatives and Industrial Efficiency 1951–64: Thirteen Wasted Years?* (London: Routledge, 1998), p. 61.
55 *Tarmac House Magazine*, Number 3 (undated).
56 The Social Class and Occupation system devised in 1911 and used in the British Census was I (professionals), II (Intermediate Occupations), III (Skilled), IV(Partly skilled),V (Unskilled).
57 For an overview, see Andrew Dainty and Martin Loosemore (eds), *Human Resource Management in Construction* (London: Routledge, 2012).

5 The Modular Society

1 R. Banham, 'Revenge of the Picturesque: English Architectural Polemics 1945–65', in John Summerson (ed.), *Concerning Architecture: Essays on Architectural Writing Presented to Nicholas Pevsner* (London: Allen Lane, 1968), p. 271.
2 Bruce Martin's papers: Statement to the Press.
3 This description appeared in the inside front cover of every edition of the *Modular Quarterly*.
4 In addition to the published output of the Modular Society, this section also draws on the archives, consisting of 14 unsorted boxes of material, deposited with the RIBA Library by Eric Corker in 1987. The author also interviewed two prominent former members of the Modular Society, Sir Roger Walters and Bruce Martin. Bruce Martin kindly made available his personal papers and records of the Technical Committee of the Modular Society and his correspondence with Mark Hartland Thomas.
5 *ModSoc News* was published between 1972 and 1976, first under the editorship of Anthony Williams and then under Bruce Martin's daughter Susan who was not replaced when she left to get married.
6 Peter Goodacre, *Cost Factors of Dimensional Co-ordination: Report of a Science Research Council Project 1976–1980* (London: Spon, 1981).
7 *Proceedings of the Ad Hoc Meeting on Standardisation and Modular Co-Ordination in Building* (Geneva: UN, 1959), Para 15.

Notes 199

8 There were 110 architect members, 11 builders, 5 government departments and 37 contractors, among other categories, in the membership cards deposited by Eric Corker. RIBA Mod.Soc. Box 15.
9 M. Hartland Thomas, 'Cheaper Building: The Contribution of Modular Co-Ordination', *Journal of the Royal Society of Arts* 101 (1953): 98–120.
10 Ibid., p. 102.
11 Ibid., p. 99.
12 *Prefabrication* 1(7) (1954): 39.
13 Bruce Martin's papers.
14 Bruce Martin's papers. Among those present were: A. Bossom, M. Blackshaw (MoHLG), D. Carter, Peter Gardiner, Ernest Hinchcliffe, H. Johnson, W. Musgrove, Monica Pidgeon, J. C. Pritchard, B. Randall, James Riley, G. Samuel, K. J. Sommerfeld, and Peter Trench. Minutes of the Preliminary Meeting of the Modular Society, mimeographed sheet.
15 Bruce Martin's papers. Minutes of the third meeting of the Provisional Committee held at the R.S.A. 1953, March 4th, 5.30 p.m.
16 RIBA Biographical file.
17 Author's private correspondence with Monica Pidgeon, 11 March 2001.
18 *AJ* (15 Jan 1959): 86–7.
19 *AR* (1936) Dec. The 14 other architects featured were: C.R. Crickmay, Mary Crowley, Gropius and Fry, Mendelsohn and Chermayeff, Oliver Hill, Frank Scarlett, E.C. Kaufmann and Benjamin, Marjorie Tall, William Lescaze, M.J.H. and Charlotte Bunney, E.C. Kaufmann, Lubetkin and Tecton, P.J.B. Hartland and Maxwell Fry.
20 *The German Building Industry*, British Intelligence Objectives Sub-Committee Trip No. 1289 (1946). Photocopy of report held at the Institute of Civil Engineer's Library.
21 Described by Gitta Sereny, in *Albert Speer: His Battle with Truth* (London: Picador, 1996), p. 30.
22 *The German Building Industry*, BIOS Trip No. 1289 (1946), p. 25.
23 Ibid., p. 26.
24 Ibid., p. 26.
25 M. Hartland Thomas, *Building Is Your Business* (London: Allan Wingate, 1947).
26 Other examples that focused on introducing industrialised building to the layman are: H. Anthony, *Houses, Permanence and Prefabrication* (London: Pleiades Books, 1945); B. Cox, *Prefabricated Houses* (London: Elek, 1945); J. Gloag and G. Wornum, *House out of Factory* (London: Allen & Unwin, 1946); R. Sheppard, *Prefabrication in Building* (London: Architectural Press, 1946).
27 *Building Is Your Business*, op. cit., pp. 70–71.
28 Ibid., p. 31.
29 ASB Study Group Number 3 (June 1949). *Dimensional Standardisation Report*, Summary and Recommendations, pt. 5.
30 'Prefabrication in Timber', *Pencil Points* (April 1943): 36–47.
31 This is described in detail in G. Herbert, *The Dream of the Factory-Made House: Walter Gropius and Konrad Wachsmann* (Cambridge, MA: MIT Press, 1984).
32 Neufert's life and work have recently been re-assessed through the publication of Walter Prigge (ed.) *Ernst Neufert. Normierte Baukultur* (Berlin: Campus, 1999).
33 A quotation from Plato ascribed to Protagoras.
34 The significance of this is examined in an essay by Thilo Hilpert, 'Menschenzeichen. Ernst Neufert und Le Corbusier', in W. Prigge (1999) op. cit.
35 *The German Building Industry* (1946) BIOS Trip No.1289, p. 25, Jean-Louis Cohen, *Architecture in Uniform: Designing and Building for the Second World War*, Canadian Centre for Architecture (Paris: Hazan, 2011), pp. 276–277.

36 *The Current Status of Modular Co-ordination*, Publication 782 (Washington, DC: National Academy of Sciences-National Research Council, 1960).
37 Mark Hartland Thomas, 'Modular Co-ordination: Hindsight and Foresight', *MQ* (1967/3): 15–21.
38 Ministry of Works, *A Survey of Prefabrication*, D. Harrison, J. M. Albery, and M.W. Whiting (London: HMSO, 1945).
39 Ibid., pp. 2, 22.
40 Ministry of Works, *Further Uses of Standards in Building*. Standards Committee Second Progress Report (London: HMSO, 1946).
41 Hartland Thomas, 'Modular Co-ordination: Hindsight and Foresight', *MQ* (1967/3): 15.
42 BRS, ed. by R. B. White (London: HMSO, 1965).
43 Ibid., p. 138.
44 The following served as members: Miss J.M. Albery, F. Austin, P.L. Locke, Anthony Cox, R. Llewelyn-Davies (chair), D. Dex Harrison, Birkin Haward, J.C. Pritchard, M. Hartland Thomas, Rodney M. Thomas, J.R. Weeks and *ex officio*, R. Sheppard and L.P. Rees.
45 ASB Study Group Number 3, *Dimensional Standardisation Report* (1949) RIBA.
46 Le Corbusier, *The Modulor: A Harmonius Measure to the Human Scale Universally Applicable to Architecture and Mechanics*. Translated by Peter de Francia and Anna Bostock (London: Faber & Faber, 1954).
47 RIBA Archives Ro/Fam/1–2 Box 2. Letter to Stamo Papadaki, New York, from Mark Hartland Thomas, 29-10-1948.
48 A summary of the research is given in M. Hartland Thomas 'Cheaper Building: The Contribution of Modular Co-ordination', *Journal of the Royal Society of Arts* 101 (1953): 98–101 and in the Technical Section, *Architect's Journal* (18 Dec. 1952): 741–747.
49 Bruce Martin's papers.
50 RIBA ASB Study Group Number 3, *Dimensional Standardisation Report* (1949) pt. 8.
51 Llewelyn-Davies expanded his ideas in 'Endless Architecture', *AA Journal* vols. 67–68, 1951–53 (November 1951).
52 This is reviewed by Eva-Marie Neumann in 'Architectural Proportion in Britain 1945–1957', *Architectural History* 39 (1995): 1977–2221. Mark Hartland Thomas entered into a long and detailed correspondence with Leonard Roberts, son of Harry Roberts, the author of a popular method taught privately as, *R's Method of Set Square Geometry and Proportion by Means of Related Set-Squares*. The method was published in a series of articles for *Architectural Design* in 1948–49, introduced by Mark Hartland Thomas. See correspondence in RIBA Archives, Ro/Fam/1–2 Box 2.
53 RIBA Archives, 'Notes for Mr Goldfinger for Modulor Meeting, 18th March 1954'. Ernö Goldfinger Papers, Box 336.
54 British Productivity Council, *Design for Production AACP Report* (1953).
55 RIBA Archives, letter from Mark Hartland Thomas to Monica Pidgeon, editor of *AD*, dated 8 June 1973. Modular Society, Box 12.
56 Hartland Thomas, 'Cheaper Building: The Contribution of Modular Co-ordination', (1953), op. cit., pp. 112–113.
57 Ibid., p. 116.
58 Ibid., p. 118.

6 'Additive architecture': the early years of modular co-ordination

1 W. R. Lethaby, *Form in Civilization: Collected Papers on Art and Labour*, 2nd edition (London: Oxford University Press, 1957), p. 166.

2 B. Martin, *School Buildings 1945–51* (London: Crosby, Lockwood and Son, 1952).
3 Patrick Geddes, *Cities in Evolution: An Introduction to the Town Planning Movement and to the Study of Civics* (London: William and Norgate, 1915); Helen Meller, *Patrick Geddes: Social Evolutionist and City Planner* (London: Routledge, 1990).
4 F.G. Novak, *Lewis Mumford and Patrick Geddes: The Correspondence* (London: Routledge 1995), p. 28.
5 Bruce Martin, *Standards in Building* (London: RIBA, 1971).
6 Bruce Martin's papers.
7 Buckminster Fuller, *Untitled Epic Poem on the History of Industrialisation* (Highlands, NJ: The Nantahala Foundation, 1962).
8 *AJ* (Dec. 18 1952). Jointing was also being studied in the private sector after Edric Neel set up ARCON in the early 1940s Rodney Thomas researched component interchangeability, jointing and modular co-ordination throughout the 1950s. The ARCON tropical bungalow was a very successful export for Taylor-Woodrow. See N. Moffett, 'Architect/Manufacturer Co-operation', *Architectural Review* 118(705) (1955): 201–204.
9 *Modular Society Transactions* 1(2).
10 Ibid., p. 3.
11 *Modular Society Transactions* 1(4).
12 Ibid., p. 4.
13 Appendix 1 gives a complete list of the buildings visited by members of the Modular Society between 1953 and 1956.
14 A. Saint, *Towards a Social Architecture: The Role of School-Building in Post-War England* (New Haven, CT: Yale University Press, 1987), p. 103.
15 *Modular Society Transactions* 1(4): 6–7.
16 Saint (1987), op. cit., pp. 105–107.
17 *AJ* (Aug. 27 1953): 265–270.
18 Ibid., p. 268.
19 Ibid., p. 269.
20 Ibid., p. 269.
21 British Standards Institute (1951) *Modular Co-ordination,* BS 1708: 1951.
22 By the mid-1960s the Modular Society was running an advisory service. This supplied consultants, who were members of its Technical Committee, to provide advice to industrial members of the Society at a fee of 25 guineas, half retained by the Modular Society and half by the consultant.
23 *MQ* (1967/3) : 18.
24 *MQ* (1956/4): 27–29.
25 Mark Hartland Thomas, 'Design for Modular Assembly of Modular and Non-Modular Components', *MQ* (1956/3): 27–29.
26 A.L. Osborne and R.A. Sefton Jenkins, 'Joints and Tolerances in Modular Structures', *MQ* (1955/3): 50; Bruce Martin, 'The Size of a Modular Component', *MQ* (1956/4): 16–23.
27 *MQ* (1959/1): 18.
28 R.A. Sefton Jenkins, 'Design for a Modular Assembly of Modular and Non-Modular Components', *MQ* (1956–57/1): 22–23.
29 *MQ* (1957/2): 24.
30 Ibid., p. 24.
31 *MQ* (1958/2): 17–18.
32 Ibid., p. 17.
33 Modular Society Archives, RIBA, Box 3, Technical Committee Minutes, 1957.
34 Ibid., Technical Committee Minutes 29-10-58. The RIBA archives contain no

information on what finally happened to the Modular Assembly once it was disassembled. Personal enquiries to ex-members have also failed to locate its final destination. Bruce Martin thought it had probably been scrapped.
35 A Founder Member of the Modular Society, Managing Director of Bovis (Contracting) Ltd. 1952–59, he became Director of the National Federation of Building Trade Employers (NFBTE) in 1959.
36 '1st Public Forum on the Modular Assembly', *MQ* (1958-9/1): 14–17.
37 Ibid., p. 16.
38 Ibid., p. 17.
39 Ibid., p. 23.
40 RIBA Archives, Modular Society Minutes of the Technical Committee, 1-12-58.
41 RIBA Archives, Modular Society Technical Committee Minutes, January 1959.
42 *AJ* (1 Jan. 1959): 9.
43 Author's personal communication, 4 March 2000.
44 *MQ* (1956/1): 16.
45 *MQ* (1964/3): 22.
46 *AJ* (8 July 1964): 69.
47 Ibid.
48 A full set of photographs and drawings appeared in *MQ* (1966/3): 6–13, including a photograph of H.R.H. Princess Margaret visiting the stand and 'inspecting' an exhibition of modular plumbing pipework.
49 *Set Square* 1(6) (1966): 6–7.
50 *AD* 34 (August 1964): 364.
51 *AD* 38 (January 1968): 3.

7 The BRS and the mathematisation of architectural modularity

1 European Productivity Agency, *Modular Co-ordination in Building: First Report of EPA Project 174* (Paris: OEEC, 1956), p. 8.
2 L. March and P. Steadman, *The Geometry of Environment* (London: RIBA, 1971), p. 222.
3 RIBA Archives, Modular Society, 'Art and Science Compared', a paper given by Mark Hartland Thomas at the RIBA on 24 February 1948.
4 Jay Hambidge, *Dynamic Symmetry: The Greek Vase* (New Haven, CT: Yale University Press, 1920); Matila Ghyka, *Le Nombre d'Or* (Paris, 1931).
5 Matila Ghyka, *The Geometry of Art and Life* (1946) and Colin Rowe, 'The Mathematics of the Ideal Villa', *AR* (March 1947): 101– 4. Much later Lionel March's 1998 book, *Architectonics of Humanism: Essays on Number in Architecture* (Chichester: Academy Editions, 1998), revisits the same theme.
6 W. A. Allen, 'Modular Co-ordination Research: The Evolving Pattern', *MQ* (1955 Summer): 14–25.
7 Ibid., p. 15.
8 Ibid., p. 16.
9 Ministry of Works, *Quicker Completion of House Interiors: Report of the Bailey Committee* (London: HMSO, 1953). This had recommended planning grids, based on preferred dimensions as an interim measure on the path to modular co-ordination.
10 E.D. Ehrenkrantz, 'Development of the Number Pattern for Modular Co-ordination: Flexibility through Standardization', *MQ* (1955/4): 39, and *MQ* (1956/1): 22–33.
11 E. Ehrenkrantz, *The Modular Number Pattern: Flexibility through Standardisation* (London: Alec Tiranti, 1956), p. 4.
12 Ibid., p. 12.
13 Ibid., p. 13.

14 Fibonacci series are generated by adding together two consecutive terms, e.g. 1, 1, 2, 3, 5 . . .
15 The back cover of Ehrenkrantz's book, published in 1956, contained a compartment holding three transparent plastic sheets, each printed with the number pattern, which could be fitted together with the four bolts and spacers provided to make a model intended to show the relationships between the numbers.
16 Ibid., p.17.
17 Ibid., p. 54.
18 RIBA Archives, MOD/SOC Box 3, *Technical Committee Minutes, 9th February 1956*.
19 Saint (1987), op. cit., p. 210.
20 Andrew Rabeneck has compared Ehrenkrantz's SCDS system for school building with that in post-war Britain in 'Building for the Future: Schools Fit for Our Children', *Construction History* 26 (2011): 55–79.
21 A. Carew, *Labour under the Marshall Plan: The Politics of Productivity and the Marketing of Management Science* (Manchester: Manchester University Press, 1987). See Chapter 11, 'The European Productivity Agency', pp. 184–201.
22 European Productivity Agency, *Modular Co-ordination. Second Report of EPA Project 174* (Paris: OEEC, 1961).
23 These were initially: Austria, Belgium, Denmark, France, Germany, Greece, Italy, the Netherlands, Norway, Sweden and the United Kingdom. The United States and Canada were associated with the project as observer countries and at a later stage Iceland, Turkey and Yugoslavia also became observer countries.
24 European Productivity Agency (1961), op. cit., p. 13.
25 Ibid., p. 30.
26 Bruce Martin's papers. Bruce Martin, 'Brick Sizes, the 4 Inch Module and Modular Brickwork,' a paper presented to the Modular Society (London: BSI, 1958).
27 *The Brick Bulletin* 1(1) (1947).
28 *The Brick Bulletin* 1 (9) (1950).
29 Bruce Martin's papers: loose insert on House of Commons notepaper.
30 Bruce Martin's records of Modular Society Technical Committee meetings.
31 *MQ* (1957/2): 14–19.
32 Ibid., p. 16.
33 Ibid., p. 19.
34 European Productivity Agency (1961), op. cit.
35 *MQ* (1959/2).
36 Bruce Martin's papers. Letter from Modular Society to BSI Technical Committee B/94, March 1960.
37 Ibid.
38 Bruce Martin's papers. Personal letter to Bruce Martin from Mark Hartland Thomas, 19 October 1961.
39 Bruce Martin's papers. Personal letter to Bruce Martin from Mark Hartland Thomas, 10 August 1962.
40 *RIBAJ* 69(7) (1962): 246.
41 Peter Goodacre, *Cost Factors of Dimensional Co-ordination: Report of a Science Research Council Project 1976–1980* (London: Spon, 1981), p. 50.

8 'Put nobody between the architect and the men': the role of architects on site

1 Transcript of interview with Alan Meikle, Hampstead, 2000.
2 A. Saint, 'The Story of CLASP', in *Towards a Social Architecture: The Role of School-Building in Post-War England* (New Haven, CT: Yale University Press,

1987), Chapter 6, pp. 157–183. CLASP was a technical system of building in lightweight steel components and also an organisational system whereby a number of local authorities pooled resources to create economies in bulk ordering of supplies.
3. Henry Swain, *RSM Bulletin* (1971), unpublished document.
4. Interview with Alan Meikle.
5. Ibid.
6. *Architectural Design* 46(5) (1976): 275–280.
7. Henry Swain, *RSM Bulletin* (1971), unpublished document which became the basis for a Technical Study published in the *Architect's Journal* as 'Notts Builds: Project RSM', *AJ* (12 January 1972).
8. In 1971, those architects who had run at least one RSM job were: Alan Willis, Roger Cheney, David Makin, Alan Reed, Gilbert Dabbous, Nick Whitehouse and Dewi Evans.
9. I am indebted to Alan Meikle for his long account of the project and the loan of his diaries, which record the project in great detail and also to Nick Whitehouse, Managing Director of Terrapin Ltd., interviewed in Milton Keynes on 24 June 2003.
10. These were all primary schools except where stated otherwise: Cotgrave, Bingham, Burton Joyce (old people's home), Arnold, Ollerton, Southwell (infant school), Eastwood (special school) and Ravenshead (Junior and Infants school).
11. G. Higgin and N. Jessop, *Communications in the Building Industry* (1963) London: Tavistock Publications, 1963), pp. 41–2.
12. *RSM Bulletin*, p. 7.
13. Interview with Alan Meikle.
14. So called because they were based on the General Motors system of drawings that showed the information required for each operation in assembling a car.
15. Henry Swain, 'Notts Builds: Project RSM' (1972) *AJ* Information Library, 12 January, p. 87.
16. They were: B. Hibbert and G. Townrow (foremen); J. Middleton, D. Hadley, K. Wilkinson, J. Noble, R. Cross, G. Paxon, and M. Savage (craftsmen joiners); B. Revitt (bricklayer); V. Callidane, R. Wass, C. Chadburn, M. Denny, C. Fox, and B. Bunting (labourers).
17. Interview with Alan Meikle.
18. This term is used in the Midlands and North of England. In the South and in London, a joiner is a bench worker, making doors, windows and other joinery items, whereas carpenters work on site; although both would have been through the same carpentry and joinery apprenticeship.
19. Neither Alan Meikle or Nick Whitehouse remembered ever encountering any union officials during the RSM project but this does not necessarily mean that the workers were not union members.
20. Interview with Alan Meikle.
21. Ibid.
22. Interview with Nick Whitehouse, Milton Keynes, June 2003.
23. Author's interview with Roger Walters.
24. Donald Gibson, 'The Needs of Our Industry and the Way Ahead,' *RIBAJ* 70(9) (1963): 361–2.
25. For example, *Dimensions and Components for Housing with Special Reference to Industrialised Building*, 1963, MPBW and the statement D.C.2 *Dimensional Co-Ordination for Industrialised Building*, 1963, MPBW followed by further details in D.C. 4,5,6 and 7.
26. *Architect and Building News* 225(13) (25 March 1964): 524.
27. Robert Elwall, *Building a Better Tomorrow* (London: RIBA, 2000), p. 17.
28. Ibid., p. 108.

29 Trevor Dannatt, *Modern Architecture in Britain* (London: Batsford, 1959), p. 42.
30 G.E. Kidder Smith, *The New Architecture of Europe* (Harmondsworth: Penguin, 1962), p. 59.
31 MQ (1955/3): 46–47.
32 Alan Diprose, 'A Review of Progress', MQ (1962/2).
33 The scheme was described in detail in MQ (1961/4): 15–22.
34 MQ (1958/2): 19–28.
35 RIBA Biographical file on Farmer and Dark.
36 Author's personal correspondence with Alan Crocker, November 2003.
37 Alan Crocker, *Module and Metric: The Theory and Practice of Dimensional Co-Ordination in Metric* (London: Pall Mall Press, 1971), p. 108.

9 The nature of work in the construction industry

1 Taken from Ultan Cowley (ed.) *McAlpine's Men: Irish Stories from the Sites* (Duncormick: Potter's Yard Press, 2010).
2 This account is proposed in K.O. Morgan, *The People's Peace* (Oxford, Clarendon Press, 1999), especially Chapter 1, 'The Façade of Unity' and also explored in depth through an analysis of the role of the black market during and after the war in Ina Zweiniger-Bargielowska's, *Austerity in Britain: Rationing Controls and Consumption* (Oxford: Oxford University Press, 2000).
3 D. V. Glass (ed.) *Social Mobility in Britain* (London: Routledge and Kegan Paul, 1954); Central Office of Information, *Social Changes in Britain* (London: HMSO, 1962); Ferdinand Zweig, *The Worker in an Affluent Society* (London: Heinemann, 1961); J. Goldthorpe, D. Lockwood, F. Bechhofer and J. Platt, *The Affluent Worker: Industrial Attitudes and Behaviour* (London: Cambridge University Press, 1968a), and J. Goldthorpe, D. Lockwood, F. Bechhofer and J. Platt, *The Affluent Worker: Political Attitudes and Behaviour* (London: Cambridge University Press, 1968b).
4 J. Goldthorpe, D. Lockwood, F. Bechhofer, and J. Platt, *The Affluent Worker in the Class Structure* (Cambridge: Cambridge University Press, 1969).
5 R. Roberts, *The Classic Slum: Salford Life in the First Quarter of the Century* (Manchester: Manchester University Press, 1971).
6 MRC. MSS. 78/BO/UM/43/2. *The Operative Builder* 2(4) (1949): 201.
7 AJ 119 (1954): 137–141.
8 MRC. MSS. 78/BO/UM/43. *Operative Builder* 1(6) (1948): 9.
9 Max Gagg, 'The Subby Bricklayer', in R Fraser (ed.) *Work*, Volume 2 (Harmondsworth: Penguin, 1969).
10 Government Social Survey, *Operatives in the Building Industry*. ed. G. Thomas (London: HMSO, 1968). A sample of 1,634 men were interviewed from both private sector firms and local authorities.
11 J.M. Sykes, 'Work Attitudes of Navvies', *Sociology* 3 (1969): 21–34.
12 See Huw Beynon's *Working for Ford* (London: Allen Lane, 1973), Dennis et al.'s study of a Yorkshire mining community, *Coal is Our Life* (London: Eyre and Spottiswoode, 1956), and Tom Lupton's experiences of an engineering workshop (*On the Shop Floor: Two Studies of Workshop Organisation and Practice*) (Oxford: Pergamon, 1963).
13 Ibid., p. 22.
14 Norah Davis, 'Attitudes to Work: A Field Study of Building Operatives', *British Journal of Psychology* 38(3) (March 1948).
15 Government Social Survey, *Operatives in the Building Industry* (1968), op. cit.
16 Cited in the *Phelps-Brown Report* (London: HMSO, 1968), p. 29.
17 For example, H. Newby, *The Deferential Worker: A Study of Farm Workers in East Anglia* (Harmondsworth: Penguin, 1977), and Dennis et al., (1956) op. cit.

18 See F. Zweig, *Labour, Life and Poverty* (London: Gollancz, 1948) and *Productivity and the Trade Unions* (Oxford: Basil Blackwell, 1951).
19 Brian Behan, *With Breast Expanded* (London: MacGibbon & Kee, 1964).
20 Ibid., p. 107.
21 Interview with Don Baldrey, Islington, May 1999.
22 K. O'Connor *The Irish in Britain* (London: Sidgewick & Jackson, 1972).
23 Ibid., p. 116. Brian Behan worked as a casual building labourer in London throughout the 1950s and 1960s. A member of the Communist Party, he was a blacklisted union activist throughout the 1960s. He was responsible for organising a mass 'go-slow' on the Festival of Britain site and was one of the organisers of the 1959 strike on the Shell site also on the South Bank. He spent time in jail after being arrested on the picket lines of both South Bank strikes.
24 Patrick MacGill, *Moleskin Joe* (1923). Republished in 1983 by Caliban.
25 Donall MacAmhlaigh, *An Irish Navvy: The Diary of an Exile* (London: Routledge & Kegan Paul, 1964), p. 150.
26 Behan (1964), op. cit.
27 MRC. MSS .78/BO/UM/43. *Operative Builder* 8(4) (1955): 247.
28 Clare Lorenz, *Women in Architecture* (London: Trefoil, 1990), p. 143.
29 Architects are anyway considered 'a load of old women' by building workers. A Skillcentre instructor made a point of informing the author, and the other trainees on her TOPS carpentry course, of this in 1978.
30 Feature on women architects in *Set Square* 1(8) (1966): 8–11.
31 M. Fogarty, *Women in the Architectural Profession* (London: Policy Studies Institute, 1978).
32 This was the conclusion of A.M. Carr-Saunders, D. Caradog Jones, and C.A. Moser, in *A Survey of Social Conditions in England and Wales* (Oxford, Clarendon Press, 1958). Here they assert that the majority of the population calls itself middle class and that occupation serves as good indicator of social class as there is 'in the public mind a fairly close association between occupation and social prestige', p. 116.
33 Anthony B. Sampson, *The Anatomy of Britain* (London: Hodder and Stoughton, 1962), p. 622.
34 *AJ* 119 (1953): 124.
35 M. Abrams, *'Architects': An Observer Survey* (London, 1964), quoted in G. Salaman, *Community and Occupation: An Exploration of Work/Leisure Relationships* (Cambridge: Cambridge University Press, 1974). Questionnaires were sent to 2,126 fellows and associates of the RIBA throughout Britain, eliciting a 64 per cent response rate; 4 per cent of respondents were women.
36 C. Gotch, 'The Architect', in R. Fraser (ed.) *Work*, Volume 2 (Harmondsworth: Penguin, 1969).
37 An early and broad overview, more anecdotal than sociological in its approach, was that of Roy Lewis and Angus Maude's, *The English Middle Classes* (London: Phoenix House, 1949).
38 G. Salaman (1974), op. cit., p. 67.
39 Ibid., p. 72.
40 Ibid., p. 78.
41 Alan Lipman, 'The Architectural Belief System and Social Behaviour', *British Journal of Sociology* 20 (1969): 190–204.
42 Social status in relation to cultural capital is discussed in P. Bourdieu's *Practical Reason: On the Theory of Action* (Cambridge: Polity Press, 1994). See Chapter 1, 'Social space and symbolic space' and Chapter 2, 'The new capital'.
43 RIBA, The Sir Percy Thomas Committee, *The Present and Future of Private Architectural Practice* (London: RIBA, 1950). In 1949, there were 3,698 architects in public service compared with 4,750 in private practice.

44 RIBA Archives. Minutes of the Salaried and Official Architects Committee (14th May 1953), Volume 1, pp. 713–714.
45 RIBA Archives, ibid.
46 RIBA Archives, ibid. For example, a series of letters dated between 1939 and 1945 from E. D. Jones, RDC Abertawe (Swansea); Samuel T. Heath, Borough Engineers Office, Wigan; H.K. Albert, Architecture and Town Planning Department, Oxford. E.D. Jones feared he would have to join the ABT in order to argue his case.
47 RIBA Archives, ibid. Letter dated May 1951 and signed by F. Jackson, H.R. Lister and others.
48 RIBA Archives, ibid. Letter dated 8th May 1951and signed by Leonard Howitt, Chairman.
49 RIBA Archives, *Report of the Sub Committee Appointed to Consider the Representation of Salaried Architects* (The Howitt Report) (London: RIBA, 1953).
50 Previously known as the Association of Architects, Surveyors and Technical Assistants (AASTA), its membership was at its highest in the 1930s and represented 'in the most clear-cut fashion the Left wing of architecture', according to John Summerson in 'Bread and Butter and Architecture', *Horizon* vol. I, pp. 234–243. AASTA changed its name to the ABT in 1942.
51 'A Trade Union for Architects? RIBA Asked To Think Again', *AJ* 12 (May 1955): 632.
52 Ian Bowen was already on the staff as editor of *The Industry* section of the *AJ*.
53 *AJ* 119 (1954): 132.
54 Transcript of author's interview with Alan Meikle, May 1999.
55 With thanks to Richard Hill for this observation.
56 Ibid.

10 Elusive connections: architects and building workers in mid-century Britain

1 RIBA, *The Architect and His Office* (London: RIBA, 1962), p. 149.
2 Department of Employment and Productivity, *British Labour Statistics. Historical Abstract 1886–1968* (London: HMSO, 1971), Tables 47 and 48.
3 *The Architect and His Office*, op. cit, p. 26.
4 BRS, *Collection of Construction Statistics* (London: HMSO, 1971), Table 7.19.
5 *RIBAJ* viii (1901): 385–399.
6 Ibid., p. 387.
7 Lethaby's ideas are described in 'Lethaby and the Fabians', Chapter 4 of Mark Swenarton's, *Artisans and Architects: The Ruskinian Tradition in Architectural Thought* (Basingstoke: Macmillan, 1989).
8 *RIBAJ* v(xxiv) (1917): 110.
9 This experiment was made, but Lethaby delegated the role of site architect on the building of All Saints Church at Brockhampton in 1902. Described in G. Rubens, *W.R. Lethaby: His Life and Work, 1857–1931* (London: Architectural Press, 1986).
10 *RIBAJ* viii (1901): 393.
11 Quoted in G. Rubens (1986), op. cit., p. 213.
12 D.A.G. Reid, Chapter IX, 'Building', in P. Venables (ed.), *Technical Education: Its Aims, Organisation and Future Development* (London: G. Bell & Son, 1955), p. 293.
13 R.W. Postgate, *The Builder's History* (London: NFBTO 1923), pp. 445–446.
14 A leader in the *Architects' Journal* for 7 October 1948, entitled 'Training Centres for Building Workers' expressed concern at the very few numbers of trainees in construction and its effect on future manpower.

15 *Architects' Journal* 110 (2835) (1949): 393–394.
16 'The Building Partnership', *AA Journal* v LXIV(732) (1949): 204–220.
17 Ibid., p. 208.
18 *The Operative Builder* 2(4) (1949): 199–202.
19 Ibid., p. 200.
20 'The Trend of Education in the Building Industry' *AA Journal* vols. 67–68 Feb. (1953) pp. 141–149.
21 'An Incomplete Team', *Architects' Journal* 117(3030) (1953): 389–340.
22 William Lethaby, *Form in Civilisation; Collected Papers on Art and Labour*, 2nd edn, Oxford: Oxford University Press, 1957.
23 Bernard Cox, *Prefabricated Houses* (London: Paul Elek, 1945), p. 31.
24 Hugh Anthony, *Houses, Permanence and Prefabrication* (London: Pleiades Books, 1945).
25 Owen Luder, 'The Future for Architects', *Set Square* 1(1) (Jan. 1966).
26 Author's interview with Alan Meikle, May 1999.
27 Ministry of Works, *Survey of Problems Before the Construction Industries* [The Emmerson Report] (London: HMSO, 1962). Section V 'Relations between building owner, professions and contractor', Pt. 27.
28 Garth Higgin and N. Jessop, *Communications in the Building Industry: A Pilot Study* (London: Tavistock Publications, 1963).
29 The entire project is reviewed by D. Boyd, and A. Wild as 'Tavistock Studies into the Building Industry: "Communications in the Building Industry 1965", and "Interdependence and Uncertainty 1996"', in M. Murray and D. Langford (eds) *Construction Reports 1944–1998*, Oxford: Blackwell, 2003).
30 Quoted in Saint, *Towards a Social Architecture*, op. cit. (1978), p. 251.
31 Description of Stirrat Johnson-Marshall by Andrew Saint in (1987), op. cit., frontispiece.
32 Interview with Sir Andrew Derbyshire recorded in 2010 for the Leverhulme Trust-funded project, 'Constructing Post-War Britain: Building Workers' Stories, 1950–70'.
33 'The Task before the Building Industry', a message from Sir Luke Fawcett, President of the NFBTO and General Secretary of the AUBTW, in *Building in Britain Today*, Ministry of Works and Central Office of Information (London: HMSO, 1949).
34 See *Building a Community: Construction Workers in Stevenage 1950–1970*, a pamphlet published as part of the Leverhulme Trust-funded project, 'Constructing Post-War Britain: Building Workers' Stories, 1950–70' and accessed via www.buildingworkerstories.com.

Bibliography

Addison, P. (1994) *The Road to 1945. British Politics and the Second World War.* London: Pimlico.
Anon (1936) *Britain Without Capitalists: A Study of What Industry in a Soviet Britain Could Achieve, by a Group of Economists, Scientists and Technicians.* London: Lawrence & Wishart.
Anthony, H. (1945) *Houses, Permanence and Prefabrication.* London: Pleiades Books.
Association of Building Technicians (1946) *Homes for the People.* London: Paul Elek.
Ball, M. (1988) *Rebuilding Construction: Economic Change in the British Construction Industry.* London: Routledge.
Banham, R. (1968) 'Revenge of the Picturesque: English Architectural Polemics 1945–65', in J. Summerson (ed.) *Concerning Architecture: Essays on Architectural Writing Presented to Nicholas Pevsner.* London: Allen Lane, p. 271.
Barham, H. (1945) *Building as a Public Service.* Dunfermline: J.B. Mackie.
Barham, H. (1947) *The Building Industry: A Criticism and Plan for the Future*, Industrial Democracy Series, No. 1. London: St. Botolph Publishing Ltd.
Barnett, C. (1986) *The Audit of War: The Illusion and Reality of Britain as a Great Nation.* London: Macmillan.
Behan, B. (1964) *With Breast Expanded.* London: MacGibbon & Kee.
Bemis, A.F. (1936) *The Evolving House*, vol. III, *Rationalisation.* Cambridge, MA: MIT Press.
Beynon, H. (1973) *Working for Ford.* London: Allen Lane.
Biernacki, R. (1995) *The Fabrication of Labor: Germany and Britain, 1640–1914.* Berkeley, CA: University of California Press.
Black, M. (1962) *The Architect's Anguish.* Leicester: Leicester University Press.
Bonnell, D.G.R. (1959) 'Operational Research in Building', in *Building Research and Documentation: The First CIB Congress.* Rotterdam: CIB.
Bossom, A. (1934) *Building to the Skies: The Romance of the Skyscraper.* London: Studio.
Bossom, A. (1944) *Post-war Building*, No. 27 in Design for Britain Series. London and Letchworth: J.M. Dent & Sons.
Bourdieu, P. (1994) *Practical Reason: On the Theory of Action.* Cambridge: Polity Press.
Bowen, I. (1951) 'The Control of Building', in *Lessons of the British War Economy.* Cambridge: Cambridge University Press.

Bowley, M. (1966) *The British Building Industry: Four Studies in Response and Resistance to Change*. Cambridge: Cambridge University Press.

Boyd, D. and Wild, A. (2003) 'Tavistock Studies into the Building Industry: "Communications in the Building Industry, 1965" and "Interdependency and Uncertainty, 1996"', in M. Murray and D. Langford (eds) *Construction Reports 1944–1998*. Oxford: Blackwell.

Braverman, H. (1974) *Labor and Monopoly Capital: The Degradation of Work in the Twentieth Century*. New York: Monthly Review Press.

Brech, E.F.L. (ed.) (1971) *Construction Management in Principle and Practice*. London: Longman.

Briar, C. (1997) *Working for Women: Gendered Work and Welfare Policies in Twentieth Century Britain*. London: UCL Press.

Brockmann, M., Clarke, L., Winch, C., Hanf, G., Méhaut, P. and Westerhuis, A. (eds) (2011) *Knowledge, Skills and Competence in the European Labour Market: What's in a Vocational Qualification?* London: Routledge.

Bruce, A. and Sandback, H.C. (1944) *A History of Prefabrication*. New York: John B Pierce Foundation.

Bullock, N. (2002) *Building the Post-War World*. London: Routledge.

Burke, P. (1992) *History and Social Theory*. Cambridge: Cambridge University Press.

Burnett, J. (1986) *A Social History of Housing 1815–1985*. London: Methuen.

Carew, A. (1987) *Labour under the Marshall Plan: The Politics of Productivity and the Marketing of Management Science*. Manchester: Manchester University Press.

Carr-Saunders, A.M., Caradog Jones, Y. and Moser, C.A. (1958) *A Survey of Social Conditions in England and Wales*. Oxford: Clarendon Press.

Clarke L. (1992) *Building Capitalism: Historical Change and the Labour Process in the Production of the Built Environment*. London: Routledge.

Clarke L. (1999) 'The Changing Structure and Significance of Apprenticeship with Special Reference to Construction', in P. Ainley and R. Rainbird (eds) *Apprenticeship*. London: Kogan Page.

Clarke, L., McGuire, C. and Wall, C. (2012) 'The Development of Building Labour in Britain', in A. Dainty (ed.) *HRM in Construction: Critical Perspectives*. London: Routledge.

Clarke, L. and Wall, C. (1997) *A Blueprint for Construction Training*. Bristol: Policy Press.

Clarke, L. and Wall, C. (2010) 'Skilled Versus Qualified Labour: The Exclusion of Women from the Construction Industry', in M. Davis (ed.) *Class and Gender in British Labour History: Renewing the Debate (or Starting it?)*. London: Merlin Press.

Clarke, L. and Winch, C. (eds) (2007) *Vocational Education: International Approaches, Developments and Systems*. London and New York: Routledge.

Cleeve Barr, A.W. (1958) *Public Authority Housing*. London: Batsford.

Coad, R. (1979) *Laing: The Biography of Sir John W. Laing CBE, 1879–1978*. London: Hodder and Stoughton.

Cohen, J-L. (2011) *Architecture in Uniform. Designing and Building for the Second World War*. Canadian Centre for Architecture. Paris: Hazan.

Connell, R.W. (1995) *Masculinities*. Cambridge: Polity Press.

Coppock, R. and Heumann, H. (1948) *Design for Labour*. No. 13 in Design for Britain Series. London and Letchworth: J.M. Dent & Sons.

Cotgrove, S. (1958) *Technical Education and Social Change*. London: George Allen & Unwin.
Cowley, U. (ed.) (2010) *McAlpine's Men: Irish Stories from the Sites*. Duncormick: Potter's Yard Press.
Cox, B. (1945) *Prefabricated Houses*. London: Paul Elek.
Crocker, A. (1971) *Module and Metric: The Theory and Practice of Dimensional Co-ordination in Metric*. London: Pall Mall Press.
Crinson, M. and Lubbock, J. (1994) *Architecture, Art or Profession? Three Hundred Years of Architectural Education in Britain*. London: Prince of Wales Institute of Architecture.
Crompton, R., Gallie, D. and Purcell, K. (eds) (1996) *Changing Forms of Employment: Organisations, Skills and Gender*. London: Routledge.
Cullingworth, J.B. (1966) *Housing and Local Government in England and Wales*. London: Allen & Unwin.
Dahrendorf, R. (1959) *Class and Class Conflict in Industrial Society*. London: Routledge.
Dainty, A. and Loosemore, M. (eds) (2012) *Human Resource Management in Construction*. London: Routledge.
Dannatt, T. (1959) *Modern Architecture in Britain*. London: Batsford.
Davenport, N. (1964) *The Split Society*. London: Gollancz.
Davies, A.J. (1992) *To Build a New Jerusalem: The British Labour Party from Keir Hardie to Tony Blair*. London: Abacus.
Dennis, N., Henriques, F. and Slaughter, C. (1956) *Coal is Our Life*. London: Eyre & Spottiswoode.
Diamant, R.M.E. (1970) *Industrialised Building*, vols 1, 2 and 3. London: ILIFFE Books.
Donnison, D.V. (1960) *Housing Policy since the War*. Welwyn: Coldicote Press.
Dunleavy, P. (1981) *The Politics of Mass Housing in Britain 1945–75*. Oxford: Clarendon Press.
Edgerton, D. (1996) *Science, Technology and the British Industrial 'Decline' 1870–1970*. Cambridge: Cambridge University Press.
Edgerton, D. (2006) *Warfare State: Britain 1920–1970*. Cambridge: Cambridge University Press.
Ehrenkrantz, E. (1956) *The Modular Number Pattern: Flexibility through Standardisation*. London: Alec Tiranti.
Ehrenkrantz, E. (1966) 'Modular Materials and Design Flexibility', in G. Kepes (ed.) *Module Symmetry Proportion*. London: Studio Vista.
Elwall, R. (2000) *Building a Better Tomorrow*. London: RIBA.
Esher, L. (1981) *A Broken Wave: The Rebuilding of England, 1940–80*. London: Allen Lane.
Finnimore, B. (1989) *Houses from the Factory: System Building and the Welfare State*. London: Rivers Oram Press.
Fogarty, M. (1978) *Women in the Architectural Profession*. London: Policy Studies Institute.
Foster, C. (1969) *Building with Men: An Analysis of Group Behaviour and Organization in a Building Firm*. London: Tavistock.
Francis, M. (1997) *Ideas and Politics under Labour 1945–51: Building a New Britain*. Manchester: Manchester University Press.
Fry, E.M. (1944) *Fine Building*. London: Faber & Faber.

Bibliography

Fuller, R. Buckminster (1962) *Untitled Epic Poem on the History of Industrialisation*. Highlands, NJ: The Nantahala Foundation.

Gagg, M. (1969) 'The Subby Bricklayer', in R. Fraser, (ed.) *Work*, volume 2. Harmondsworth: Penguin.

Gann, D. (2000) *Building Innovation*. London: Thomas Telford.

Geddes, P. (1915) *Cities in Evolution: An Introduction to the Town Planning Movement and to the Study of Civics*. London: William & Norgate.

Ghyka, M. (1931) *Le Nombre d'Or*, Paris: Sheed & Ward, republished Dover, 1977.

Ghyka, M. (1946) *The Geometry of Art and Life*. New York: Sheed and Ward.

Gilbreth, F. (1908) *Concrete System*. New York: Engineering News Publishing Company.

Gilbreth, F. (1909) *Bricklaying System*. New York: Myron C. Clark.

Glass, D. V. (ed.) (1954) *Social Mobility in Britain*. London: Routledge & Kegan Paul.

Glendinning, M. and Muthesius S. (1994) *Tower Block: Modern Public Housing in England, Scotland, Wales, and Northern Ireland*. New Haven, CT: Yale University Press.

Gloag, J. and Wornum, G. (1946) *House out of Factory*. London: Allen & Unwin.

Goldthorpe, J., Lockwood, D., Bechhofer, F. and Platt, J. (1968a) *The Affluent Worker: Industrial Attitudes and Behaviour*. London: Cambridge University Press.

Goldthorpe, J., Lockwood, D., Bechhofer, F. and Platt, J. (1968b) *The Affluent Worker: Political Attitudes and Behaviour*. London: Cambridge University Press.

Goldthorpe, J., Lockwood, D., Bechhofer, F. and Platt, J. (1969) *The Affluent Worker in the Class Structure*. Cambridge: Cambridge University Press.

Goodacre, P. (1981) *Cost Factors of Dimensional Co-ordination: Report of a Science Research Council Project, 1976–1980*. London: Spon.

Gotch, C. (1969) 'The Architect', in R. Fraser (ed.) *Work*, vol. 2. Harmondsworth: Penguin.

Gourvish, T. and Tiratsoo, N. (ed.)(1998) *Missionaries and Managers: American Influences on European Management Education, 1945–60*. Manchester: Manchester University Press.

Hall, D. (1948) *Cornerstone: A Study of Britain's Building Industry*. London: Lawrence & Wishart.

Hall, N. (1965) *Report of the Joint Committee on Training in the Building Industry*. London: RIBA.

Hambidge, J. (1920) *Dynamic Symmetry: The Greek Vase*. New Haven, CT: Yale University Press.

Hanson, B. (2003) *Architects and the 'Building World' from Chambers to Ruskin: Constructing Authority*. Cambridge: Cambridge University Press.

Hardless, T. (ed.) (1969) *Europrefab Systems Handbook*. London: Interbuild Prefabrication Publications Ltd.

Hennessy, P. (1993) *Never Again: Britain 1945–51*. London: Vintage.

Herbert, G. (1978) *Pioneers of Prefabrication: The British Contribution in the Nineteenth Century*. Baltimore, MD: The Johns Hopkins University Press.

Herbert G. (1984) *The Dream of the Factory-Made House: Walter Gropius and Konrad Wachsmann*. Cambridge, MA: MIT Press.

Higgin, G. and Jessop, N. (1963) *Communications in the Building Industry – Report of a Pilot Study*. London: Tavistock Publications.

Higgin, G. and Jessop, N. (1965) *Communications in the Building Industry*. London: Tavistock Publications.
Hillebrandt, P. (1984) *Analysis of the British Construction Industry*. London: Macmillan.
Hilton, W.S. (1963) *Foes to Tyranny: A History of the AUBTW*. London: AUBTW.
Hilton, W.S. (1968) *Industrial Relations in Construction*. Oxford: Pergamon.
Hoggart, R. (1957) *The Uses of Literacy*. London: Pelican.
Jackson, A. (1970) *The Politics of Architecture: A History of Modern Architecture in Britain*. London: Architectural Press.
Jackson, T.G. and Shaw, N. (eds) (1892) *Architecture: An Art or Profession*. London: John Murray.
Jenkins, A. (1971) *On Site 1921–71*. London: Heinemann.
Kaye, B. (1960) *The Development of the Architectural Profession in Britain: A Sociological Study*. London: George Allen & Unwin.
Kidder Smith, G.E. (1962) *The New Architecture of Europe*. Harmondsworth: Penguin.
Kidson, P., Murray, P. and Thompson, P. (1965) *A History of English Architecture*. Harmondsworth: Penguin.
Knox, F. and Hennessy, J. (1966) *Restrictive Practices in the Building Industry*. London: The Institute of Economic Affairs.
Landes, D. (1969) *The Unbound Prometheus: Technological Change and Industrial Development in Western Europe from 1750 to the present*. Cambridge: Cambridge University Press.
Landes, D. (1998) *The Wealth and Poverty of Nations*. London: Little Brown and Company.
Layton, E. (1961) *Building by Local Authorities. The Report of an Enquiry by the Royal Institute of Public Administration into the Organisation of Building and Construction by Local Authorities in England and Wales*. London: George Allen & Unwin.
Layton, E. (1962) *Report on the Practical Training of Architects*. London: RIBA.
Le Corbusier (1954) *The Modulor: A Harmonius Measure to the Human Scale Universally Applicable to Architecture and Mechanics*. Trans. P. de Francia and A. Bostock. London: Faber & Faber.
Leon, G. (1971) *The Economics and Management of System Construction*. London: Longman.
Lethaby, W. (1957) *Form in Civilisation: Collected Papers on Art and Labour*, 2nd edn. Oxford: Oxford University Press.
Lewis, R. and Maude, A. (1949) *The English Middle Classes*. London: Phoenix House.
Lockwood, D. (1958) *The Blackcoated Worker*. London: Allen & Unwin.
Lorenz, C. (1990) *Women in Architecture*. London: Trefoil.
Lupton, T. (1963) *On the Shop Floor: Two Studies of Workshop Organisation and Practice*. Oxford: Pergamon
MacAmhlaigh, D. (1964) *An Irish Navvy: The Diary of an Exile*. London: Routledge & Kegan Paul.
MacGill, P. (1983) *Moleskin Joe*. London: Caliban Press. Originally published 1923, Herbert Jenkins.
March, L. (1998) *Architectonics of Humanism: Essays on Number in Architecture*. Chichester: Academy Editions.

March, L. and Steadman, P. (1971) *The Geometry of Environment*. London: RIBA.
Maré de, E. (ed.) (1948) *New Ways of Building*. London: Architectural Press.
Martin, B. (1952) *School Buildings 1945–51*. London: Crosby, Lockwood & Son.
Martin, B. (ed.) (1965) *The Dimensional Co-ordination of Building*. London: RIBA.
Martin, B. (1971) *Standards in Building*. London: RIBA.
Martin, L. and March, L. (1972) *Urban Space and Structures*. Cambridge: Cambridge University Press.
Marwick, A. (1968) *Britain in the Century of Total War*. London: Bodley Head.
McKellar, E. (1999) *The Birth of Modern London: The Development and Design of the City*. Manchester: Manchester University Press.
McKibbin, R. (1998) *Classes and Cultures in England, 1918–1950*. Oxford: Oxford University Press.
Meller, H. (1990) *Patrick Geddes: Social Evolutionist and City Planner*. London: Routledge.
Merrett, S. (1979) *State Housing in Britain*. London: Routledge & Kegan Paul.
Miller, M. (1992) *Raymond Unwin: Garden Cities and Town Planning*. Leicester: Leicester University Press.
Morgan, K.O. (1984) *Labour in Power, 1945–51*. Oxford: Oxford University Press.
Morgan, K.O. (1999) *The People's Peace: British History, 1945–1990*. Oxford: Clarendon Press.
Morris, W. (1973) *Political Writings of William Morris*, ed. A.L. Morton. London: Lawrence & Wishart.
Mowery, D. (1986) 'Industrial Research in Britain 1900–1950', in D.B. Elkun and W. Lazonick (eds) *The Decline in the British Economy*. Oxford: Clarendon Press.
Neufert, E. (1936) *Bauentwurfslehre*. Berlin: Gütersloh.
Neufert, E. (1943) *Bauordnungslehre*. Berlin: Bauwelt-verlag.
Newby, H. (1977) *The Deferential Worker: A Study of Farm Workers in East Anglia*. Harmondsworth: Penguin.
Novak, F.G. (1995) *Lewis Mumford and Patrick Geddes: The Correspondence*. London: Routledge.
Ockmann. J. (1993) *Architecture Culture 1943–68*. New York: Rizzoli.
O'Connor, K. (1972) *The Irish in Britain*. London: Sedgewick & Jackson.
Peters, T.F. (1996) *Building the Nineteenth Century*. Cambridge, Massachusetts, London: MIT Press.
Perry, P.J.C. (1976) *The Evolution of British Manpower Policy: From the Statute of Artificers 1563 to the Industrial Training Act 1964*. Chichester: the author.
Pevsner, N. (1960) *Pioneers of Modern Design*. Harmondsworth: Penguin Books.
Postgate, R.W. (1923) *The Builder's History*. London: The National Federation of Building Trade Operatives.
Powell, C. (1980) *An Economic History of the British Building Industry 1815–1979*. London: Architectural Press.
Powers, A. (1993) 'Arts and Crafts to Monumental Classicism: The Institutionalisation of Architectural Classicism 1900–1914', in N. Bingham (ed.) *TheEducation of the Architect: Proceedings of the 22nd Annual Symposium of theSociety of Architectural Historians of Great Britain*. London: Society of Architectural Historians, pp. 34–38.
Preziosi, D. (1983) *Minoan Architectural Design: Formation and Significance*. Berlin: Mouton.

Price, R. (1997) 'Postmodernism as Theory and History', in J. Belchem and N. Kirk (eds) *Languages of Labour*. London: Ashgate.

Prigge, W. (ed.) (1999) *Ernst Neufert. Normierte Baukultur*. Berlin: Campus.

Pugin, A.W.N. (1836)*Contrasts, or a parallel between the noble edifices of the fourteenth and fifteenth centuries and similar buildings of the present day; shewing the present decay of taste*. London: A.W.N. Pugin.

Pugin, A.W.N. (1841)*The True Principles of Pointed or Christian Architecture: set forth in two lectures delivered at St. Marie's, Oscott*. London: J. Weale.

Reid, D.A.G. (1955) 'Building', in P. Venables (ed.) *Technical Education, its Aims, Organisation and Future Development*. London: G. Bell & Sons.

Roberts, R. (1971) *The Classic Slum: Salford Life in the First Quarter of the Century*. Manchester: Manchester University Press.

Rosenberg, N. (1960) *Economic Planning in the British Building Industry 1945–49*. Philadelphia, PA: University of Philadelphia Press.

Rubens, G. (1986) *W.R. Lethaby: His Life and Work 1857–1931*. London: Architectural Press.

Ruskin, J. (1892) 'On the Nature of Gothic Architecture: and Herein of the True Functions of the Workman in Art'. Reprinted from the sixth chapter of the second volume of Mr. Ruskin's *Stones of Venice*. London: G. Allen.

Ruskin, J. (1904) *The Crown of Wild Olive*. London: George Allen.

Ruskin, J. (1910) *The Seven Lamps of Architecture, The Lamp of Life XX1V*. London: George Routledge & Sons Ltd. Ruskin, J. (2000) *Fors Clavigera. Letters to the Workmen and Labourers of Great Britain*, (ed.) D. Birch. Edinburgh: Edinburgh University Press.

Ruskin, J. (2004) 'Selected Writings' in D. Birch (ed.) *The Stones of Venice, II, 1853, The Nature of Gothic*. Oxford: Oxford University Press, p. 47.

Russell, B. (1981) *Building Systems, Industrialization and Architecture*. London: John Wiley & Sons, Ltd.

Saint, A. (1987) *Towards a Social Architecture: The Role of School-Building in Post-War England*. New Haven, CT: Yale University Press.

Saint, A. (2007) *Architect and Engineer: A Study in Sibling Rivalry*. London: Yale University Press.

Salaman, G. (1974) *Community and Occupation: An Exploration of Work/Leisure Relationships*. London: Cambridge University Press.

Sampson, A. (1962) *The Anatomy of Britain*. London: Hodder & Stoughton.

Sanderson, M. (1994) *The Missing Stratum: Technical School Education in England 1900–1990s*. London: The Athlone Press.

Satoh, A. (1995) *Building in Britain: The Origins of a Modern Industry*. Aldershot: Scolar Press.

Savage, M. (2000) 'Sociology, Class and Male Manual Work Cultures', in I. Campbell, N. Fishman, and J. McIlroy (eds) *British Trade Unions and Industrial Politics*, volume II, *The High Tide of Industrial Trade Unionism, 1964–79*. Aldershot: Ashgate, pp. 23–41

Schnaidt, C. (1965) *Hannes Meyer: Buildings, Projects and Writings*. London: A. Tiranti.

Schwartz, F.J. (1996) *The Werkbund and Design Theory. Mass Culture Before the First World War*. New Haven, CT: Yale University Press.

Sebestyén, G. (1998) *Construction: Craft to Industry*. London: E. and F.N. Spon.

Sennett, R. (2003) *Respect: The Formation of Character in a World of Inequality*. London: Penguin, Allen Lane.
Sennett, R. (2008) *The Craftsman*. London: Allen Lane.
Sereny, G. (1996) *Albert Speer: His Battle with Truth*. London: Picador.
Sheppard, R. (1946) *Prefabrication in Building*. London: Architectural Press.
Simon, E.D. (1945) *Re-building Britain: A Twenty Year Plan*. London: Gollancz.
Sissons, M. and French, P. (eds) (1963) *The Age of Austerity*. London: Hodder & Stoughton.
Skedd, A. and Cook, C. (1984) *Post-war Britain: A Political History*. Harmondsworth: Penguin.
Steadman, P. and March, L. (1971) *The Geometry of Environment: An Introduction to spatial Organization in Design*. London: RIBA.
Stone, P. (1963) *Housing, Town Development and Land Costs*. London: Estates Gazette.
Stone, P.A. (1966) *Building Economy: Design, Production and Organisation: A Synoptic View*. Oxford: Pergamon Press.
Summerson J. (1991) *Georgian London*. London: Pimlico.
Swain, Henry (1971) RSM Bulletin, unpublished document.
Swenarton, M. (1989) *Artisans and Architects: The Ruskinian Tradition in Architectural Thought*. Basingstoke: Macmillan.
Taylor, F.W. and Thompson, S. (1912) *Concrete Costs: Tables and Recommendationsfor Estimating the Time and Cost of Labor Operations in Concrete Construction and for Introducing Economical Methods of Management*. New York: John Wiley.
Thomas, M.H. (1947) *Building Is Your Business*. London: Allan Wingate.
Thompson, E.P (1980) *The Making of the English Working Class*. London: Gollancz.
Thurow, L. (1974) *Generating Inequality: Mechanisms of Distribution in the U.S. Economy*. New York: Macmillan.
Tomlinson, J. (1991) 'The Labour Government and the Trade Unions, 1945–51', in N. Tiratsoo (ed.) *The Attlee Years*. London: Pinter.
Tomlinson, J. (1994) 'The Politics of Economic Measurement: The "Productivity Problem" in the 1940s', in A. Hopwood and P. Miller (eds) *Accounting as Socialand Institutional Practice*. Cambridge: Cambridge University Press.
Tomlinson, J. and Tiratsoo, N. (1998) *The Conservatives and Industrial Efficiency 1951–64. Thirteen Wasted Years?* London: Routledge.
Tuckman, A. (1982) 'Looking Backwards – Historical Specificity of the Labour Process in Construction', in *BISS Proceedings*. London: UCL Press.
United Nations (1974) *Dimensional Co-ordination in Building: Current Trends and Policies in EEC Countries*. New York: United Nations.
Unrau, J. (1981) 'Ruskin, the Workman and the Savageness of Gothic', in R. Hewison (ed.) *New Approaches to Ruskin*. London: Routledge, pp. 33–50.
Ure, A. (1835) *The Philosophy of Manufactures: or an Exposition of the Scientific, Moral and Commercial Economy of the Factory System of Great Britain*. London: Charles Knight.
Wachsmann, K. (1961) *The Turning Point of Building*. Trans. Thomas E. Burton. New York: Rheinhold.
Wallis, L. (1945) *The Building Industry: Its Work and Organisation*. London: J. M. Dent & Sons.
Weaver, H.J.O. (1962) 'Innovation and the Operatives', in *Supplement to Innovation*

in Building: Contributions at the Second CIB Congress, Cambridge. Amsterdam: Elsevier Publishing Company.
Whittick, A. (1943) *Civic Design and the Home*, No. 10 in Rebuilding Britain Series. London: Faber.
Wiener, M. (1981) *English Culture and the Decline of the Industrial Spirit*. Cambridge: Cambridge University Press.
Williams-Ellis, Clough (1941) *Plan for Living: The Architect's Part*, No. 5 in Rebuilding Britain Series. London: Faber.
Wood, L. (1979) *A Union to Build: The Story of UCATT*. London: Lawrence & Wishart.
Wood, S. (ed.) (1989) *The Transformation of Work?: Skill, Flexibility and the Labour Process*. London: Unwin Hyman.
Woodhuysen, J. and Abley, I. (eds) (2004) *Why Is Construction So Backward?* Chichester: Wiley.
Wright, E.O. (1997) 'Rethinking, Once Again, the Concept of Class Structure', in J. Hall (ed.) *Reworking Class*. Ithaca, NY: Cornell University Press.
Zweig, F. (1948) *Labour, Life and Poverty*. London: Gollancz.
Zweig, F. (1951) *Productivity and the Trade Unions*. Oxford: Basil Blackwell.
Zweig, F. (1961) *The Worker in an Affluent Society: Family, Life and Industry*. London: Heinemann.
Zweiniger-Bargielowska, I. (2000) *Austerity in Britain: Rationing Controls and Consumption*. Oxford: Oxford University Press.

Government and Official Publications

British Intelligence Objectives Sub-Committee Trip No. 1289, (1946) *The German Building Industry*.
British Productivity Council (1950) *AACP Report on Building*. Anglo-American Productivity Council.
British Productivity Council (1953) *Design for Production. AACP Report*. Anglo-American Productivity Council.
British Productivity Council (1954) *Getting Together: A Review of Productivity in the Building Industry*. British Productivity Council.
British Productivity Council (1959) *Better Methods in the Building Trades*. British Productivity Council.
British Standards Institute (1951) *Modular Co-ordination*. BS 1708: 1951.
Building Apprenticeship and Training Council (1943) *First Report*. London: HMSO.
Building Apprenticeship and Training Council (1944) *Second Report*. London: HMSO.
Building Apprenticeship and Training Council (1949) *Fourth Report*. London: HMSO.
Building Apprenticeship and Training Council (1956) *Final Report*. London: HMSO.
Building Research Station (1944a) *No. 1 House Construction*. Post-War Building Studies. London: HMSO.
Building Research Station (1944b) *No.2 Standard Construction for Schools*. Post-War Building Studies. London: HMSO.
Building Research Station (1944c) *No. 8 Reinforced Concrete Structures*. Post-War Building Studies. London: HMSO.

218 Bibliography

Building Research Station (1965) *Prefabrication: A History of its Development in Great Britain*, ed. R.B. White. London: HMSO.

Building Research Station (1966a) *Architects and Productivity*, ed. D. Bishop, BRS Current Papers, Design Series 57. London: HMSO.

Building Research Station (1966b) *Building Operatives' Work,* vols 1 and 2, ed. R.E. Jeanes. London: HMSO.

Building Research Station (1971) *Collection of Construction Statistics.* London: HMSO.

Central Office of Information (1962) *Social Changes in Britain.* London: HMSO.

CITB (1966) *Report and Statement of Accounts.* London: HMSO.

CITB (1967) *Report and Statement of Accounts.* London: HMSO.

CITB (1969) *A Plan of Training for Operative Skills in the Construction Industry.* London: HMSO.

Cmnd. 939 (1960) *The Royal Commission on Doctors' and Dentists' Remuneration.* London: HMSO.

Construction Industry Training Board (1966) *Report and Statement of Accounts.* London: HMSO.

Department of Employment and Productivity (1968) *Report of the Committee of Enquiry into Certain Matters Concerning Labour in Building and Civil Engineering*, (Phelphs-Brown Report). London: HMSO.

Department of Employment and Productivity (1970a) *Managing Men in Construction.* London: HMSO.

Department of Employment and Productivity (1970b) *Men on Site.* London: HMSO.

Department of Employment and Productivity (1971) *British Labour Statistics. Historical Abstract 1886–1968.* London: HMSO.

Department of the Environment (1980) Construction Statistics Division, *Statistics Collected by the Ministry of Works 1941–56*, compiled by M.C. Fleming. London: HMSO.

Department of the Environment (1994) *Constructing the Team: Joint Review of Procurement and Contractual Arrangements in the UK Construction Industry*, ed. M. Latham. London: HMSO.

Department of the Environment/Construction Industry Board, IPRA Ltd. and University of Westminster (1997) *Strategic Review of Construction Training.* London: Thomas Telford.

Department of the Environment, Transport and the Regions (1998) *Rethinking Construction: The Report of the Construction Task Force.* London: DETR.

Department of Health (1924) *New Methods of House Construction: Interim Report.* London: HMSO.

DETR/DoE (1999) *Innovative Manufacturing Initiative*, Cabinet Office Technology Foresight Programme. London: HMSO.

European Productivity Agency (1956) *Modular Co-ordination in Building, Project Number 174.* Paris: OEEC.

European Productivity Agency (1961) *Modular Co-ordination, Second Report of EPA Project 174.* Paris: OEEC.

Government Social Survey (1968) *Operatives in the Building Industry,* ed. G. Thomas. London: HMSO.

Harris, A.I. (1966) *Labour Mobility in Great Britain 1953–63*, Government Social Survey Report, No. SS 333 (March). London: HMSO.

Hoagland, A. (1998) 'The Invariable Model. Standardisation and Military

Architecture in Wyoming 1860–1900', *Journal of the Society of Architectural Historians* 57(3): 298–315.

International Council for Building Research (1961) *Building Research and Documentation: Contributions and Discussions at the First CIB Congress, Rotterdam, 1959*. Amsterdam: Elsevier.

Kohan, C.M. (1952) *Works and Buildings*, History of the Second World War United Kingdom Civil Series, ed. W.K. Hancock. London: HMSO.

Local Government Board (1918) *Report of the Committee Appointed . . . to Consider Questions of Building Construction in Connection with the Provision of Dwellings for the Working Classes etc*, (The Tudor Walters Report), Cmnd. 9191. London: HMSO.

MHLG (1965) *Industrialised Housebuilding*, Circular No. 76/65. London: HMSO.

Ministry of Health (1948) *The Cost of House-building*, Girdwood Committee's First Report. London: HMSO.

Ministry of Housing and Local Government (1963) *Dimensional Co-ordination in Housebuilding*, Circular No. 59/63. London: HMSO.

Ministry of Housing and Local Government (1965) *Housing Programme 1965–70*, Cmnd. 2838. London: HMSO.

Ministry of Labour (1965) *Manpower Studies No. 3. Construction Industry*. London: HMSO.

Ministry of Labour and National Service (1951) *Welfare Arrangements on Building Sites*. London: Ministry of Labour and National Service.

Ministry of Public Building and Works (1963) *A National Building Agency*, Cmnd. 2228. London: HMSO.

Ministry of Public Building and Works (1966) *Conference Report: Economic and Social Research in Construction*. London: HMSO.

Ministry of Works (1942) *Report on Training for the Building Industry*, Central Council for Works and Buildings, Education Committee. London: HMSO.

Ministry of Works (1943) *Training for the Building Industry*, Cmnd. 6428. London: HMSO.

Ministry of Works (1944) *Methods of Building in the USA*. London: HMSO.

Ministry of Works (1945) *A Survey of Prefabrication*, ed. D. Harrison, J.M. Albery, and M.W. Whiting. London: HMSO.

Ministry of Works (1946) *Further Uses of Standards in Building*, Standards Committee, Second Progress Report. London: HMSO.

Ministry of Works (1947a) *Payment by Results in the Building and Civil Engineering Industries*. London: HMSO.

Ministry of Works (1947b) *Summary Reports of the Ministry of Works*, Cmnd. 7279. London: HMSO.

Ministry of Works (1948) *Summary Reports of the Ministry of Works*, Cmnd. 7541. London: HMSO.

Ministry of Works (1949) *Summary Reports of the Ministry of Works*, Cmnd. 7698. London: HMSO.

Ministry of Works (1950a) *Summary Reports of the Ministry of Works*, Cmnd. 7995. London: HMSO.

Ministry of Works (1950b) *Working Party Report on Building*. London: HMSO.

Ministry of Works (1950c) *Machines for the Modern Builder*. London: HMSO.

Ministry of Works (1950d) *Wages, Earnings and Negotiating Machinery in the Building Industry since 1886*, Chief Scientific Adviser's Division, Economics Research Section. London: HMSO.

Ministry of Works (1951) *Summary Reports of the Ministry of Works*, Cmnd. 8306. London: HMSO.

Ministry of Works (1953) *Quicker Completion of House Interiors: Report of the Bailey Committee*. London: HMSO.

Ministry of Works (1962) *Survey of Problems before the Construction Industries*. Report Prepared for the Minister of Works by Sir Harold Emmerson. London: HMSO.

Ministry of Works, National Building Studies (1948) *New Methods of House Construction*, Special Report No. 4. London: HMSO.

Ministry of Works, National Building Studies (1949) *New Methods of House Construction*, Second Report, Report No. 10. London: HMSO.

Ministry of Works, National Building Studies (1950) *Productivity in House-building: A Pilot Sample Survey in the South, East and West of England and in South Wales August 1947–Oct. 1948*, Special Report No. 18. London: HMSO.

Ministry of Works, National Building Studies (1953) *Productivity in House-building*, Second Report, Special Report No. 21, eds W.J. Reiners and H.F. Broughton. London: HMSO.

Ministry of Works, National Building Studies (1958) *Incentives in the Building Industry*, Report No. 28. eds Alison Entwhistle and W.J. Reiners. London: HMSO.

Ministry of Works, National Building Studies (1959a) *Organisation of Building Sites EPA Project No. 302/1*, Report No. 29, ed. R.C. Sansom. London: HMSO.

Ministry of Works, National Building Studies (1959b) *A Study of Alternative Methods of House Construction*, Report No. 30, eds W.J. Reiners and H.F. Broughton. London: HMSO.

Ministry of Works, National Building Studies (1960) *Mobile Tower Cranes for Two and Three Storey Building*, Report No. 31, eds H.F. Broughton, W.J. Reiners and H.G. Vallings. London: HMSO.

National Academy of Sciences–National Research Council (1960) *The Current Status of Modular Co-ordination*, Publication 782. Washington: USA.

National Building Agency (1970) *Industrialised 2-Storey House Building. A Productivity Study*. London: National Building Agency.

OEEC (1954) *Cost Savings Through Standardization, Simplification, Specialisation in the Building Industry*. Paris: OEEC.

OEEC (1959) *Proceedings of the Ad Hoc Meeting on Standardisation and Modular Co-Ordination in Building*. Geneva: UN. RIBA (1943) *Report of the Special Committee on Architectural Education*. London: RIBA.

RIBA (1949) *Dimensional Standardisation Report*, ASB Study Group Number 3. London: RIBA.

RIBA (1950) *The Present and Future of Private Architectural Practice*, The Sir Percy Thomas Committee. London: RIBA.

RIBA Archives (1953) *Report of the Sub Committee Appointed to Consider the Representation of Salaried Architects* (The Howitt Report).London: RIBA.

RIBA (1962) *The Architect and his Office*. London: RIBA.

UN (1970) *Modular Co-Ordination of Low-Cost Housing*. New York: United Nations.

UN (1974) *Dimensional Co-Ordination in Building: Current Trends and Policies in EEC Countries*. New York: United Nations.

Archives

MRC MSS. 78/BO/UM/4/1/18-40. *NFBTO Annual Conference Reports*, 1947–1969. London: National Federation of Building Trades Operatives.

MRC MSS. 78/BO/UM/43/1-14. *The Operative Builder*, volumes 1–14 1947–1961. London: National Federation of Building Trades Operatives.

MRC MSS. 78/BO/UM/4/2/3. April 1959, Report of Conference on New Techniques in the Building Trades. MRC MSS. 78/BO/UM/4/1/123.

TNA PRO CAB 21/2026. Minutes of the Cabinet Production Committee, 31 March 1950.

TNA PRO ED 150/45. Letter from Michael Waterhouse, President of the RIBA to G. Tomlinson, Minister of Education, 13 January 1950.

TNA PRO ED 150/45. Minute Sheet, 18 January 1950.

TNA PRO LAB 8/424. Quoted in *Notes from a Meeting of the Hankey Technical Personnel Committee* (undated).

TNA PRO LAB 8/464. *Memorandum of War Time Agreement on Employment of Women in the Building Industry 1941–42*.

TNA PRO LAB 8/518. *The Post-War Demand for Labour in the Building and Civil Engineering Industries* by G.D.H. Cole.

TNA PRO LAB 8/867. *Designated Craftsmen Scheme*. Letter from T.A. Barlow, Treasury Chamber, to Hugh Beaver, Director-General, Ministry of Works.

TNA PRO LAB 8/867. Letter to Gould at the Ministry of Labour and National Service.

TNA PRO LAB 8/867. Letter to Voysey, Ministry of Works from London Master Builder's Association, 30 July 1943.

TNA PRO LAB 8/1095. *Post-war Reconstruction: Control of Numbers of Architects Needed*.

TNA PRO LAB 8/1095. Letter from E.G. McAlpine, Ministry of Education to Miss B. Grainger, Ministry of Labour and National Service, 5 Feb. 1947.

TNA PRO LAB 8/1095. Central Council for Works and Buildings, Training Sub Committee Minutes, 31 May 1943.

TUC HD 6661 W8 *Amalgamated Society of Woodworkers Monthly Journal*, 2 (Feb. 1942): 83.

TUC HD 6661 NFBTO *The Operative Builder, 1953–60*, B9.

TUC HD 9715/6 NFBTO Annual Conference, 1945.

TUC LRD 2/B/Building *New Builder's Leader and Electricians Journal*, 8(1943): 8.

TUC LRD 2/B/1 *New Builder's Leader*, 6(6) (May 1941): 24.

TUC LRD 2/B/1/Building *New Builder's Leader*, January 1945 (10)4.

Journal articles and theses

Allen, W.A. (1955) 'Modular Co-ordination Research: The Evolving Pattern', *MQ* 1955/2: 14–25.

Bartlett International Summer School Proceedings (BISS) (1979–1992). London: University College London.

Bennett, T.P. (1943) 'The Architect and the Organisation of Post-War Building', *RIBAJ* Jan.: 63–74.

Binney, H.A., Walters, R. and Weston, J. (contributors) (1967) 'Standardisation: The Way Ahead', Report of seminar for the building industry organised by the Modular Society at Royal Society of Arts, October 1967, *The Modular Quarterly* 4: 10–27.

Brockman, M., Clarke, L. and Winch, C. (2008) 'Competence-based Vocational Education and Training (VET) in Europe: The Cases of England and France'. *Vocations and Learning* 1(3): 227–244.

Bullock, N. (2005) 'Re-assessing the Post-War Housing Achievement: The Impact of War-damage Repairs on the New Housing Programme in London', *Twentieth Century British History* 16(3): 256–282.

Burchell, S. (1979) 'Training: An Analysis of the Crisis', in *BISS Proceedings*, UCL.

Burgess, R.A. (1968)'Accuracy and Productivity in IB', *Building,* 18 November, 189–98.

Caponella, A. and Vittorini, R. (2003) 'Building Practices of the Post-War Reconstruction Period in Italy: Housing by Mario Ridolfi at the INA Casa Tiburtino Neighbourhood in Rome 1950–54', in *Proceedings of the First International Congress on Construction History*, Madrid, January 2003, ed. Santiago Huerta, vol.1, Madrid.

Clarke, L. and Janssen, J. (2008) 'Forum: A Historical Context for Theories Underpinning the Production of the Built Environment', *Building Research and Information* 36(6), 659–662.

Clarke, L. and Wall, C. (2010) ' "A Woman's Place is Where She Wants to Work": Barriers to the Retention of Women in the Building Industry after the Second World War', *Scottish Labour History:* 16–39.

Clarke L. and Winch C. (2006) 'A European Skills Framework? – But What Are Skills? Anglo-Saxon Versus German Concepts', *Journal of Education and Work* 19(3): 255–269.

Cole, G.D.H. (1942) 'The Labour Problem in the Building Industry', *Agenda. A Quarterly Journal of Reconstruction* 1: 129–143.

Cole, G.D.H. (1943) 'The Building Industry after the War', *Agenda. A Quarterly Journal of Reconstruction* 2: 146–152.

Davis,N. (1948) 'Attitudes to Work: A Field Study of Building Operatives', *British Journal of Psychology* 38(3)March.

Diprose, A. (1962) 'A Review of Progress', *MQ* (2).

Ehrenkrantz, E.D. (1955) 'Development of the Number Pattern for Modular Coordination: Flexibility through Standardization,' *MQ* 1955/4 and *MQ* 1956/1.

Gibson, D. (1963)'The Needs of Our Industry and the Way Ahead,' *RIBAJ* 70(9): 361-2.

Green, A. (2002) 'Poor Relation or "The General Education of the Working Class": VET in a Historical Perspective', paper presented at ESRC seminar series, 'Vocational Education: Historical Developments, Aims and Values', University of Leicester, June.

Harper, D. (1964) 'Some Notes on the Education of the Architect and his Technical Assistant', *RIBAJ* 71(10): 441–442.

Hayes, N. (2002) 'Did Manual Workers Want Industrial Welfare? Canteens, Latrines and Masculinity on British Building Sites 1918–1970', *Journal of Social History* Spring: 637–658.

Jones, H. (2000) '"This is Magnificent": 300,000 Houses a Year and the Tory Revival after 1945', *Contemporary British History* 14(1): 99–121.
Knight, A.V. (1956) 'A Critique of Work Study', *Work Study and Industrial Engineering* 3, December, p. 81.
Lipman, A. (1969) 'The Architectural Belief System and Social Behaviour', *British Journal of Sociology* 20: 190–204.
Llewelyn-Davies, R. (1951) 'Endless Architecture', *AA Journal* 67–68, 1951–53: 106–113.
Llewelyn-Davies, R. and Weeks, J. (1962) 'Educating for Building', Paper Number 3, *BASA Second Report on Architectural Education, Building for People*. London: BASA.
Luder, O. (1966) 'The Future for Architects', *Set Square* 1(1) Jan.
Martin, B. (1956) 'The Size of a Modular Component', *MQ* 1956/4: 16–23.
Mason, J. (1946) 'A Report on Structural Engineering in Germany', *The Structural Engineer* June, p. 334.
McCormick, B.J. (1964) 'Trade Union Reaction to Technological Change in the Construction Industry', *Yorkshire Bulletin of Economic and Social Research* 16(1): 15–30.
McGuire, C., Clarke, L. and Wall, C. (2013) 'Battle on the Barbican: The Struggle for Trade Unionism in the British Building Industry, 1965–7', *History Workshop Journal* (in press).
Moffett, N. (1955)'Architect/Manufacturer Co-operation', *Architectural Review* 18(705): 201–204.
Neumann, E.-M. (1995)'Architectural Proportion in Britain 1945–1957', *Architectural History* 39: 1977–2221.
Osborne, A.L. and Sefton Jenkins, R.A. (1955) 'Joints and Tolerances in Modular Structures', *MQ* 1955/3: 50.
Prais, S. and Steedman, H. (1986) 'Vocational Training in France and Britain: The Building Trades', *National Institute of Economic and Social Research (NIESR) Review* (May) No. 116.
Prefabrication (1954) 1(7): 39.
'Prefabrication in America', (1942) *Architect's Journal* Oct. 8, pp 231–233.
Rabeneck, A. (2011)'Building for the Future: Schools Fit for Our Children', *Construction History* 26: 55–79.
Rowe, C. (1947) 'The Mathematics of the Ideal Villa, Palladio and Corbusier Compared', *AR* (March): 101–104.
Sefton Jenkins, R.A. (1956) 'Design for a Modular Assembly of Modular and Non-Modular Components', *MQ* 1956–57/1: 22–23.
Summerson, J. (1942)'Bread and Butter and Architecture', *Horizon* vol. I, pp. 234–243.
Swain, H. (1972) 'Notts Builds: Project RSM', *AJ* 155(2): 75–96.
Sykes, J.M. (1969) 'Work Attitudes of Navvies', *Sociology* 3: 21–34.
Thomas, M. Hartland (1953) 'Cheaper Building: the Contribution of Modular Coordination', *Journal of the Royal Society of Arts* 101: 98–120.
Thomas, M. Hartland (1956) 'Design for Modular Assembly of Modular and Non-Modular Components', *MQ* 1956/3: 27–29.
Thomas, M. Hartland (1967) 'Modular Co-ordination: Hindsight and Foresight', *MQ* 1967/3: 15–21.

Thompson, E.P. (1978) 'Eighteenth-Century English Society: Class Struggle Without Class?' *Social History* 3(2): 156.

Tomlinson, J. (1991) 'The Failure of the Anglo-American Council on Productivity', *Business History* 33(1): 83–92.

Tuckman, A. (1982) 'Looking Backwards: Historical Specificity of the Labour Process in Construction', *BISS Proceedings*, UCL.

Walters, R.T. (1957) 'Towards Industrialised Building', *RIBAJ* February: 150–153.

Weeks, J. (1963–64) 'Indeterminate Architecture', *Transactions of the Bartlett Society*, 2: 85–106.

Wilson, A.R. (1966) 'The Home and the Goals of Industrialisation', Cambridge PhD 5644.

Illustrations credits

Figure 1.1   Cover of *New Ways of Building* (1948) edited by Eric de Maré, London: Architectural Press.
Figure 1.2   Trade advertisement for system building, early 1960s. Source: Stramit Technology Holdings Ltd.
Figure 2.1   Building workers c.1943 listening to a Colonel returned from the Malta campaign. Source: Imperial War Museum Photograph Archive.
Figure 2.2   Building workers c. 1943 in conversation with a Colonel returned from the Malt a campaign. Source: Imperial War Museum Photograph Archive.
Figure 2.3   Woman bricklayer in World War II. Source: Imperial War Museum Photograph Archive.
Figure 4.1   Modern plant for brickwork. Source: *New Ways of Building* (1948).
Figure 5.1   Cover of The Modular Society flyer c. mid 1960s. Source: Bruce Martin's papers.
Figure 5.2   Lord Bossom examining a Modular Society questionnaire while on a site visit. Source: *Modular Quarterly*, Autumn 1960.
Figure 6.1   Cover of *The Modular Quarterly* showing the first Modular Assembly. Source: *MQ*, Autumn 1958.
Figure 6.2   Presentation drawings for the first Modular Assembly. Source: *MQ* Autumn 1958.
Figure 6.3   Mark Hartland Thomas on the steps of the first Modular Assembly. Source: RIBA Photograph Collection.
Figure 6.4   The five essentials of modular co-ordination, a summary of five years' work. Source: *MQ*, Summer 1958.
Figure 6.5   The second Modular Assembly at IBSAC 1964. Source: *MQ*, Autumn 1964.
Figure 6.6   The second Modular Assembly exhibition of modular components. Source: *MQ*, Autumn 1964.
Figure 6.7   The third Modular Assembly at IBSAC 1966. Source: *MQ*, 1966.
Figure 6.8   Modular Society members at the Society's conference, 18 November 1963. From left to right: Alan Diprose (Modular Primer), P.H Dunstone (Combinations), Mark Hartland Thomas (Secretary), Peter Trench (Chair), Bruce Martin (International Work). Source: *MQ*, Winter 1963–64.

226 *Illustration credits*

Figure 7.1   Three-dimensional model of Ehrenkrantz's number pattern found inside the back cover of *The Modular Number Pattern* (1956). Source: Cambridge University Department of Architecture Library.

Figure 7.2   Photo of British Standards Institute test buildings, EPA Project, designed by Bruce Martin. Source: European Productivity Agency (1961) *Modular Co-ordination, Second Report of EPA Project 174*. Paris: OEEC.

Figure 7.3   Scale drawing of British Standards Institute test buildings, EPA Project, designed by Bruce Martin. Plans based on four-inch module. Source: European Productivity Agency (1961) *Modular Co-ordination, Second Report of EPA Project 174*. Paris: OEEC.

Figure 7.4   Building Research Station test buildings, EPA Project. Rationalised traditional terraced housing in brick and block. Source: European Productivity Agency (1961) *Modular Co-ordination, Second Report of EPA Project 174*. Paris: OEEC.

Figure 7.5   Plans of BRS terraced housing in brick and block. Source: European Productivity Agency (1961) *Modular Co-ordination, Second Report of EPA Project 174*. Paris: OEEC.

Figure 8.1   Little Aden Cantonment, Farmer and Dark, 1961. Photographs of site model. Source: *Modular Quarterly*.

Figure 8.2   Little Aden Cantonment, plans and elevations of officers' housing. Source: *Modular Quarterly*.

Figure 8.3   Little Aden Cantonment, plans and elevations of sergeants' mess. Source: *Modular Quarterly*.

Figure 10.1 Ratings of 'contribution to the building process' and 'social status' by occupation. Source: Redrawn from Higgins and Jessop's *Communications in the Building Industry: Pilot Study*, London: Tavistock Institute, 1963.

Figure 10.2 A75 Metric building under construction. Source: *Modular Quarterly*, 1967, No. 2.

Index

Aalto, Alvar 95–6
Abercrombie, Patrick 19
Abrams, Mark 163, 165
Ad Hoc Report on Architectural Education (1952) 51
Adshead, Stanley 19
Advisory Committee of the Building and Civil Engineering Industries 33
aesthetics 116–17
Air Raid Protection (ARP) 45
Albery, Jessica 92
Allen, William 52, 54, 117–18, 122, 124–5, 130
Amalgamated Society of Woodworkers (ASW) 30, 37, 74
Amalgamated Union of Building Trade Workers (AUBTW) 33, 41, 44, 63, 74, 181
American Institute of Architects 27
American Standards Association 82–3, 89
Anglo-American Council on Productivity (AACP) 61–2, 66, 67, 75, 93, 121–2
Anthony, Hugh 177
Appleyard, Major-General 43
apprenticeships 29–30, 41, 57–8, 59, 176–7; Cole's Report 47, 48, 49; as obstacle to industrialisation 71; post-war reconstruction 21; *Training for the Building Industry* 49; wages 73, 74
architects 1, 2–3; class position 14, 75, 162–6; communication with operatives 141–3; earnings 164, 171; industrialisation 26–8, 79, 182; modular co-ordination 91, 93, 97; Modular Society 79, 81–2, 83; post-war reconstruction 22, 181; productivity 70–1; relationships with operatives 171–80; role of 135–53; RSM project 136–7, 138, 139, 140–1, 145–6; salaried 166–70; social status 165–6; training 51–6, 60, 66, 174; United States 63; wartime period 44–7; women 161–2; work dissatisfaction 165
Architects' Journal (AJ) 47, 84, 109, 113, 141, 144, 155, 168, 174, 176
Architects Registration Act (1931) 2
Architectural Association (AA) 95, 148, 174, 175
Architectural Design (AD) 45, 84, 90–1, 107, 115
architectural determinism 165
Architectural Science Board (ASB) 87, 91–2, 186
architectural technicians 53, 55, 138
Arts and Crafts 9, 10
Arup, Ove 45
Aslin, C.H. 52, 150
Association of Building Technicians (ABT) 40, 47, 168, 171, 177
Atkinson, George 130

Baldrey, Don 159
Ball, Michael 4
Balmain, W.A. 97–8
Banham, Reyner 79
Barham, Harry 43
Barnett, C. 3
Barr, Cleeve 27
Barton, John 150
Beaver, Hugh 33
Beddington, Nadine 161
Behan, Brian 158, 159–60
Bemis, Alfred Farwell 20, 22
Bennett, John 81, 139
Bernal, J.D. 42

Bevan, Aneurin 43–4
Biernacki, Richard 11–12
Bishop, Donald 51, 70
Black, Misha 26
bonuses 139–40
Bossom, Alfred 20, 27, 67, 79, 83–4, 85, 123, 125
Bovis 41
Bowen, Ian 168
Bowley, Marian 2–3, 5
Braverman, H. 10–11
The Brick Bulletin 123
bricklayers 13, 36, 59, 72, 156, 182; RSM project 138, 144; training 58; wages 73; women 37, 39
brickwork 68; EPA Project 174 122–3, 124–5; modular 107–8; module size 126–8, 131
British Empire 19
British Intelligence Objectives Sub-Committee (BIOS) 86, 88, 89
British Productivity Council 62
British Standards Institute (BSI): EPA Project 174 122, 125; modular co-ordination 83, 93, 97, 101, 115, 119; module size 27, 92, 98, 126, 130–1
Brixton School of Building 173, 176
Buckminster Fuller, Richard 97
Building Apprenticeship and Training Council (BATC) 49, 56–8
Building Industry National Council (BINC) 33, 36, 171
building industry 30, 50, 154–70, 171; concept of skill 7–14; criticism of 177–9; experience of work 158–60; manual workers 155–6; nationalisation proposals 42–4; salaried architects 166–70; status 156–8; technical change 2–7; transformations 1; wages 74; wartime period 32–40; women 161–2
Building Industry Communications Research Project (BICRP) 179
Building Research Station (BRS) 11, 21, 157–8; criticism of 44; modular co-ordination 27, 90, 101, 117–18; module size 92, 98, 125–31; productivity 23, 69, 70; proportion 116; training 58
building workers 1, 182; attitudes 67, 75; class position 75; Cole's Report 47–8, 49; concept of skill 7–14; experience of work 158–60; Little Aden Cantonment 153; manual working class 155–6; post-war reconstruction 21, 22; productivity 64–5; relationships with architects 171–80; RSM project 137–46; Ruskin on 8–9; shortages of 35; site communication 141–3; skilled 10, 30, 35–7, 69, 71–4, 75; status 156–8; training 54, 55, 56–60, 66; unskilled 35–6, 41–2, 49, 69, 72, 73, 75; women 35–41, 161–2; *see also* labour; labourers; wages
Built Environment 79–81
Burt, George 67

Campaign Against Building Industry Nationalisation (CABIN) 44
Carew, A. 62
carpenters 30, 36, 59, 72, 172; RSM project 138; school-building programme 99–100; training 58; wages 73
Carter, David 105, 125
Carter, John 137
Casson, Hugh 92
cast iron 19
city planning 96
civil engineering 30, 50, 178–9, 180; wages 73, 74; wartime period 32–3; *see also* engineers
cladding 148
Clarke, Linda 7–8
CLASP building system 7, 103, 135–8, 139, 140, 141, 143–4, 147
class divisions 14, 20, 75, 154, 155, 178; architects 162–6; professional bodies 56; women 162; *see also* hierarchies
Clerks of Works 146, 169, 173
Codes of Practice 21
Cohen, Jean-Louis 89
Cole, G.D.H. 42–3, 47–8, 49, 56, 59
collective bargaining 61
communication 71, 141–3, 146
competition 4
concrete: GRID Method 112; housing 19; post-war reconstruction 21, 23, 25–6; RSM project 144; training 57; union concerns 30; wages 74; wartime building industry 41
Construction Industry Training Board (CITB) 58, 59–60, 70

Constructional Engineering Union (CEU) 40
consultation 29, 75
contracting system 4
contracts 70–1, 74
Cooke, Alistair 64
Coppock, Richard 19, 28–9, 36, 40, 43–4, 50, 60, 73–4, 173–4
costs: modular co-ordination 97–8; RSM project 138–9, 140
Council of Industrial Design (COID) 86, 92, 186
Cox, Anthony 168
Cox, Bernard 177
craftsmanship 28, 29, 36–7, 171–2; Cole's Report 47; Germany 9–10; Ruskin on 8, 10; training 58; wages 74; see also skill
Crinson, M. 52
Cripps, Stafford 61
Crocker, Alan 45, 150–3
Crowley, Mary 95
Crystal Palace 19
Cutbush, P. 124–5

Dannatt, Trevor 147
decorators 72
Denison, P.A. 118
Derbyshire, Andrew 147–8, 181
Design and Industries Association (DIA) 86
de-skilling 6, 7, 10, 66, 71
determinism, architectural 165
Detroit Sheets 140
Deutscher Werkbund 12, 19–20
dimensional co-ordination (dc) 1, 28, 147; definition of 188; efficiency 71; Germany 117–18; industrialisation 182; Marchwood Power Station 148; mathematisation of 116–31; Modular Society 79–94; terminology 27; see also modular co-ordination
Diprose, Alan 105, 112–13, 115, 148
Direct Labour Organisations (DLOs) 7, 136, 160
Directorate of Building Materials 21
Directorate of Post-War Building 21
Dunican, Peter 155
Dunleavy, P. 6
Dunstone, P.H. 115
Durbin, Evan 61

Edgerton, David 3, 4

education 1, 3–4, 48, 51–60; see also training
efficiency 3, 19, 23, 71
Ehrenkrantz, Ezra 118–21, 125, 130
electricians 72
Emmerson Report (1962) 54, 178–9
employment exchanges 32
engineers 2–3, 166, 180; see also civil engineering
Essential Work Order (EWO) 33
European Co-operation Administration (ECA) 61
European Productivity Agency (EPA) 116, 121–31

factories 12, 18
'families of components' 100
Farmer and Dark 147, 149, 150, 153
Fawcett, Luke 33, 43, 50, 181
Festival of Britain 73, 84, 92–3, 159
Finnimore, Brian 6
Ford, Henry 19, 20
foremen 70, 136–7, 140, 142, 159, 169, 178
fragmentation 4
Fraser, Donald 97, 98, 125
freedom 156–7, 158

Gardiner, Peter 107, 118
Geddes, Patrick 96–7
gender issues 35–41, 161–2
Germany: Cologne Cathedral 9–10; concept of labour 12–13; *Deutscher Werkbund* 12, 19–20; EPA Project 174 122; Ministry of Works Team assessment 84, 86–7; modular co-ordination 86–9, 117–18; productivity 69; textile industry 11, 12
Ghyka, Matila 117
Gibbs, Alexander 45
Gibson, Donald 4, 70–1, 135, 146–7
glaziers 59
Glendinning, Miles 6–7
Golden Mean 119
Goldfinger, Ernö 45, 92
Goldthorpe, J. 154, 158
Goodacre, Peter 131
Gotch, Christopher 163–4
Gothic architecture 8, 9
Government Training Centres (GTCs) 57
Greater London Council 22
GRID Method 112–13

Index

grid planning 82, 89–90, 93–4, 130, 188; Germany 86–7; Little Aden Cantonment 150; Summerswood School 98; terminology 27
Gropius, Walter 20, 53, 88

Hall, Sir Noel 54, 55
Hambidge, Jay 117
Hankey Report (1949) 46–7
Hanson, Brian 9
Harper, D. 54
Harrison, D. Dex 92
Hartland Thomas, Mark 27, 79–81, 82–6, 90–1, 115; disagreement with Martin 100–1; Ehrenkrantz's number pattern 121; Festival of Britain 92–3; German system 87–8; letter from 186–7; Modular Assemblies 102–3, 105, 106–7, 108, 109, 111; modular brickwork 107–8; module size 92, 97, 98, 101, 124, 130; proportion 116–17
Hatchett, Michael 110–11
Haward, Birkin 45
Henderson, Bill 150
Heumann, Harry 28, 44
Hicks, George 19, 50
hierarchies 45–6, 74–5, 155, 164, 168–9; *see also* class divisions
high-rise buildings 6–7, 25
Hillebrandt, P. 4
Hilton, W.S. 44
Hinchcliffe, Ernest 93–4
Holford, William 45
Houses of Parliament 8
housing 19, 21–6, 29, 91, 176
Howitt, Leonard 167
human scale 82

industrial relations 65
Industrial Training Act (1964) 59
Industrial Training Boards (ITBs) 59
industrialisation 1, 6, 17–31, 42, 71, 177, 181, 182; architects and 26–8, 79, 166; benefits to the workforce 137; definitions of 17, 23, 25; fragmented nature of 115; impact on the workplace 158; innovation 2; modular co-ordination 116; Modular Society 79, 81–2; pace of 67–9; policy 5; post-war reconstruction 21–6; trade unions 28–31; wartime projects 45; women workers 161

Industrialised Building Systems and Components (IBSAC) 23; Modular Assemblies 112–15
innovation 2, 3, 4
Institute of Builders 54
Institute of Civil Engineers (ICE) 57
Institute of Structural Engineers (ISE) 54, 57
Institution of Municipal and County Engineers 57
international co-operation 116
Isaacs, George 41

Jackson, Frank 37, 41
Johnson-Marshall, Stirrat 52, 91, 95, 97, 146, 179, 181
joiners 13, 30, 59, 72; RSM project 143–4; training 58; wages 73
jointing 102–3

Keystone 40

labour: availability of 32; concept of skill 7–14; efficiency 3; exploitation of 4; German conception of 12–13; 'labour power' 11, 12, 13; modular co-ordination 97–8; women 35–41; *see also* building workers; productivity; wages
labourers 13, 35–6, 37, 38–40, 41, 72, 153; Cole's Report 47–8; experience of work 159; manual working class 155–6; RSM project 138; status 156–8; wages 29, 73, 74; women 37, 40; *see also* building workers
Lacey, Dan 98, 135
Laing 26, 35, 41, 160, 176
Lalor, Oliver 154
Landes, D. 3–4
Laurence, G. 125
Layton, Elizabeth 53, 54
Le Corbusier 88, 90–1, 92, 93, 118
Lethaby, William 51, 54, 95, 136, 137–8, 171–4, 177, 178
licensing 22, 64, 66
'light and dry' construction 22, 28, 49–50, 91, 147, 178
Lipman, Alan 165
Little Aden Cantonment 147, 148–53
Llewelyn-Davies, Richard 52, 55, 56, 86, 90, 92, 93
local government 7, 167, 180
Lockwood, D. 154
London 18, 22

London Master Builders' Association 37
Lubbock, J. 52
Luder, Owen 177–8
Lyons, Eric 150

MacAmhlaigh, Donall 160
MacGill, Patrick 160
MacMorran Report (1955) 51–2
management 66–7, 70, 74–5, 137, 146, 159
Manchester Building Guild 173–4
Manzoni, Herbert 93
March, Lionel 116
Marchwood Power Station 147–8
Markus, Tom 108
MARS group 86
Marshall Plan 61, 62
Martin, Bruce 91, 95–7, 98–100, 115, 147; EPA Project 174 121, 122, 123, 124–5; Little Aden Cantonment 148; Modular Assemblies 108, 113; module size 92, 101–2, 130; *OAP* editorial board 81
Martin, Leslie 52
Marx, Karl 7, 11
Marxist perspectives 4, 10–11
masons 59, 72, 172
mass-production 19, 82, 90, 165, 181
Matthew, Robert 52, 63
Maurice, F.D. 10
McAlpine, E.G. 46
McKechnie, John 63
mechanisation 5, 10, 17, 71–3; human scale 82; Little Aden Cantonment 153; post-war reconstruction 25; prefabrication 111; Second World War 35; wartime building industry 41
Medd, David 45, 91, 95–6, 118
meetings 145
Meikle, Alan 46, 135–7, 140, 141–3, 144–6, 168–9, 178
Meyer, Montagu 43
middle class 154, 162–3, 164, 165, 178
Mills, Edward 54
Ministry of Housing and Local Government (MHLG) 23, 25, 32, 147
Ministry of Labour 32, 40–1, 46
Ministry of Supply 21
Ministry of Works (MoW): assessment of Germany 84, 86–7; engineers 166; modular co-ordination 89–90,

97; National Building Corporation proposal 43; National Buildings Record 45; post-war reconstruction 21; productivity 63; wartime building industry 32–3, 35, 36–7
Modern Movement 6, 20, 86
Modular Assemblies 102–8, 112–15
modular co-ordination 1, 2, 28, 165; definition of 188; early years of 95–115; Ehrenkrantz's number pattern 118–21; EPA Project 174 121–31; five essentials of 110; industrialisation 17, 23, 182; Little Aden Cantonment 148–53; mathematisation of 116–31; Modular Society 79–94; terminology 27; *see also* dimensional co-ordination
Modular Quarterly (MQ) 79, 102–5, 109, 111, 113, 118, 125, 148
Modular Society 2, 27–8, 79–94, 97–115, 144, 182; Ehrenkrantz's number pattern 118–21; letter from Hartland Thomas 186–7; Little Aden Cantonment 150; module size 117, 118, 119, 123, 130–1; Pavilion 112–13; proportion 116; visits to modular buildings 184–5
module, definition of 100–1
module size 27, 90–2, 97–8, 101–2, 118; Bemis 20; BRS studies 117, 125–31; Ehrenkrantz's number pattern 118–21; EPA Project 174 122–5; Germany 87, 88; human scale 82; Imperial measurement 83; Little Aden Cantonment 150; United States 82–3, 89
Morris, Henry 139
Morris, William 10
Morrison and Partners 112
multi-storey buildings 5
Mumford, Lewis 96, 177
Muthesius, Stephan 6–7

National Building Corporation proposal 43–4
National Building Studies Reports 70
National Buildings Record 45
National Consultative Council of the Building and Civil Engineering Industries 33
National Federation of Building Trade Employers (NFBTE) 56, 63, 70, 111–12, 176

232 *Index*

National Federation of Building Trade Operatives (NFBTO) 6, 28–31, 33, 36, 56, 171, 175, 177; nationalisation proposals 44, 174; wages 73–4; women workers 40, 161
National Federation of Clay Industries 123
National Joint Consultative Council of Architects, Quantity Surveyors and Builders (NJCC) 54, 179
National Joint Council for the Building Industry (NJCBI) 28, 40, 57, 58, 73
nationalisation 42–4, 57, 170, 174, 175
navvies 157–8, 160
NENK method 4, 22, 28, 147
Neufert, Ernst 86–7, 88–9
New Architecture Movement (NAM) 182
New Builder's Leader (NBL) 37–9, 41
New Pattern of Operative Training (NPOT) 59–60
New Towns 158, 177, 182
'number pattern' 118–21

Oddie, Guy 118
Official Architects' Association 166
Official Architecture and Planning (OAP) 79–81
operational research 23
The Operative Builder 37–8, 41, 65, 155, 175–6
operatives *see* building workers
output measurement 13

painters 9, 59, 72, 73
payment by results (PBR) 33, 50, 66, 88
Pevsner, Nikolaus 147
Phelps-Brown Report (1968) 61
Pidgeon, Monica 115
piece-rates 12, 13
Pierce, S. Rowland 90
Pilkington 108
Pinion, J.T. 150
plasterers 59, 72, 73, 172
plumbers 58, 59, 72, 73, 138
policy 5, 22, 181
Portal, Lord 33, 67
Postgate, R.W. 174
post-war reconstruction 1, 21–6, 42–4, 46, 50, 177, 181–2
Powell, C. 4
Powers, Alan 51

prefabrication 5, 6, 17–19, 20, 181; CLASP building system 135–6; export of prefabricated houses 83; Marchwood Power Station 148; modular co-ordination 89, 93–4, 111, 117; post-war reconstruction 21, 177; schools 99–100; threat to the brick industry 123; unskilled workers 49; wages 74; War Office 147
'process' 23
productivity 6, 27, 61–9, 75; EPA Project 174 121–2; management and 69–71; operational research 23; post-war reconstruction 21–3; RSM project 140
project management 145
proportion 92, 116–17, 119
Pugin, Augustus 8

quantity surveyors 3, 13; RSM project 138, 139, 145–6; training 54, 55

rationalisation 1, 81, 82, 90, 119, 150, 177
reconstruction 1, 21–6, 42–4, 46, 50, 177, 181–2
Reid, D.A.G. 173
Reilly, Charles 19, 22
Research into Site Management (RSM) 135–46, 160, 178
Rippon, Geoffrey 146, 147
Roberts, Robert 155
Rowe, Colin 117
Rowntree, Diana 115
Royal Institute of British Architects (RIBA) 26, 33, 164, 173, 176; BATC 57; failures of 165, 167; industrialisation 177; letter from Hartland Thomas 186–7; Modular Assembly 105; modular co-ordination 27, 87, 90, 91, 97, 130; opposition to industrialisation 182; productivity 70; salaried architects 166; training 51–5; union membership 168; wartime period 45, 46; women architects 161–2
Royal Institute of Chartered Surveyors (RICS) 54, 57, 176
Ruskin, John 8–10, 181
Russell, B. 7

Saint, A. 21
Salaman, Graeme 164–5

salaried architects 166–70
Sampson, Anthony 163
Samuely, Felix J. 86, 97
Sandys, Duncan 33
Schinkel, Karl Friedrich 9–10
school building programme 5, 6, 7, 28, 95, 99–100, 147; A75 prefabricated timber system 150; dimensional co-ordination 91; RSM project 135–7
Schwartz, F.J. 19–20
scientific management 19, 62, 69
Scott, George Gilbert 9–10
Second World War 1, 21, 32–42, 49–50, 177; architects 44–7, 170; class divisions 154; consultation during the 29; *see also* post-war reconstruction
Sefton Jenkins, R.A. 103, 125
self-employment 4, 35, 71, 141
Sennett, Richard 11
Simon, E.D. 32, 36
Simon Committee Report (1942) 48
site communication 141–3
site organisation 19, 20, 25
site welfare 155–6
Skarum, Ernst 125
skill 40–2, 69, 71–4, 174; architects 165; concept of 7–14; prefabrication 177; RSM project 143–6; skill shortages 7, 22, 144, 177, 181; *see also* craftsmanship
skilled workers 10, 30, 35–7, 69, 71–4, 75
skyscrapers 20
slaters 59, 73
Smith, G. Kidder 147
'social architecture' 137
Soviet Union 182
Speer, Albert 86
Spence, Basil 176
Standard Method of Measurement (SMM) 139, 140, 145–6
standardisation 12, 19, 20, 181; architectural details 71; EPA Project 174 125; Germany 86–7, 88; Little Aden Cantonment 150; modular co-ordination 90, 93; Modular Society 27; post-war reconstruction 21; trade unions' support for 31; United States 67
Steadman, Philip 116
steel 19, 21
Stokes, Richard 30, 66–7

structural engineers 54, 55
subcontracting 4, 58, 138, 180
Summerswood School 98–9
surveyors 3, 13; RSM project 138, 139, 145–6; training 54, 55
Swain, Henry 135, 137–8, 139–40, 141, 143–4, 169, 178
Sykes, J.M. 157

Tatchell, Sydney 90
Tatton-Brown, William 91, 98, 111
Taylor, F.W. 19
technical education 3–4
technology 177
terminology 27
Tewson, Vincent 65
textile industry 11–12
Thomas, Hugh 163
tilers 59, 73
timber 19, 21
tolerances 71, 107–9, 111–13, 119, 130
Tomlinson, George 52
Tomlinson, J. 62
trade unions 6, 28–31, 33, 181; AACP 62; architects 168; collective bargaining 61; criticism of the Building Industry Productivity Team report 65–6; industrialisation 182; Lethaby's view of 173; nationalisation proposals 42–3, 44; training 57–8, 59; wages 73–4; women workers 37, 40, 161
Trade Unions Congress (TUC) 62, 168
training 3, 51–60, 66, 174, 175; Cole's Report 47–8, 49; foremen 70; joint 54, 173, 176; management 70; new occupations 71–3; post-war reconstruction 42; problems with 3–4; Simon Committee Report 48; vocational education and training 1, 11, 12–13, 56, 176; wages 74
Training for the Building Industry (White Paper, 1943) 47, 48–9
Treibel, W. 69
Trench, Peter 23, 107, 115
Tuckman, Alan 23
The Tudor Walters Report (1918) 19

unemployment 65–6
United States: modular co-ordination 27, 82–3, 89, 187; productivity 61–2, 63–6, 67; School Construction System Development programme 121

unskilled workers 35–6, 41–2, 49, 69, 72, 73, 75
urbanisation 8
Ure, Andrew 177, 178

Vitrolite 108
vocational education and training (VET) 1, 11, 12–13, 56, 176

Wachsmann, Konrad 20, 88, 93
wages 6, 29, 32–3, 71–4, 171, 172; collective bargaining 61; impact on productivity 70; piece-rates 12, 13; RSM project 139–40; Second World War 36; United States 64; women 40, 161
Wagner, Martin 20
Wallis, Lesley 36
Walters, Roger 17, 22, 98, 118, 146
War Office 4, 146–53
Waterhouse, Michael 63

Wates 35
Weate, John 150
Weaver, Harry 29–30, 31, 174–5
Weeks, John 55
Welfare State 4, 6, 154, 163, 180, 181, 182
West, James 67
White, R.B. 5
Whitehouse, Nick 145
Wiener, M. 3
Williams, Anthony 95, 98
Wilson, Harold 4, 23, 25
Wimpey 25, 26, 35, 41
Wolstencroft, Frank 43, 67
women 35–41, 145, 161–2
Wood, L. 28
Woodbine Parish, D.E. 176
working class 154, 155–6, 162–3
working conditions 155–6, 158–60

Zweig, Ferdinand 28, 158